SCIENCE
AND
SOCIOLOGICAL PRACTICE

SCIENCE
AND
SOCIOLOGICAL PRACTICE

Steven Yearley

Department of Social Studies
The Queen's University, Belfast

OPEN UNIVERSITY PRESS
Milton Keynes

Open University Press
A division of
Open University Educational Enterprises Limited
12 Cofferidge Close
Stony Stratford
Milton Keynes MK11 1BY, England

First published 1984

British Library Cataloguing in Publication Data

Yearley, Steven
Science and sociological practice.
1. Science—Social aspects
I. Title
306′.45 Q175.5

ISBN 0-335-10584-X
ISBN 0-335-10581-5 pbk

Text and cover design by Clark Williams

Typeset by Butler & Tanner Ltd, Frome, Somerset

Printed in Great Britain by St. Edmundsbury Press, Bury St. Edmunds, Suffolk

Contents

Preface

Although this book draws widely on the philosophy, sociology and history of natural science its subject is essentially the philosophy of social science. At one level it is intended as an introduction to the many issues associated with the question of whether social science should be regarded as a science. As an introduction it should therefore be accessible to all students of sociology already familiar with the principal approaches to sociological analysis. However, unlike many books on the philosophy of social science the majority of the discussion is not about questions of principle alone. Instead the emphasis is on the practical implications which these philosophical questions have – whether for the performance of sociological research or for the evaluation of other people's contributions to the social scientific literature. Rather than asking whether, in principle, sociology can be scientific it is proposed that it is necessary to inquire what prospects there are for a practically attainable science of society. The book is thus not intended simply for those concentrating on the philosophy of social science; it is directed at least as much to those concerned with sociological theory or empirical social science.

I should like to take this opportunity of thanking Michael Mulkay and Jonathan Potter for their careful reading of the text and for their many helpful recommendations. While their suggestions did much to improve it, the book and its shortcomings remain my responsibility.

Introduction

The issue of the scientific standing of social science has long been a matter of dispute. External questioners have sought to discover how serious a science sociology can hope to be in order to decide whether sociologists' claims are worth attending to. For their part sociologists have often wanted to know whether their work is fundamentally science-like so that they can determine if they should base their techniques on those of natural science or, alternatively, begin to look elsewhere for methodological guidance.

In both these cases the question of the meaning of the term 'science' in social science is far from being scholastic. The aim of this book is to review the debate between naturalism (the belief that sociology should approach and seek to understand its subject matter in the same manner as the natural sciences) and anti-naturalism in order to establish in what methodologically significant ways sociology can be modelled on or approximate to the procedures of natural science. This is not, therefore, a philosophical undertaking in the strictest sense because the emphasis will consistently be placed on those areas of the debate about naturalism which have consequences for the direction and assessment of sociological research. Other areas of abstract philosophical interest will be comparatively neglected.

Nor will this study consist in an historical review of the arguments concerning the scientific standing of what were generically termed the 'moral sciences'. Such a review, extending back to the eighteenth century, say, might well contain many episodes with implications for the present situation of social science – as the few studies of this sort appear to indicate.[1] However, there is sufficient complexity and diversity in extant approaches to justify treatment of them in isolation. Moreover, these extant approaches are likely to underlie the empirical social studies which contemporary sociologists most often carry out, encounter or have practically to evaluate.

The arrangement of this book does reflect historical developments in one sense: three basic positions on the naturalism issue will be identified and they derive not abstractly but from general perspectives espoused by groups of practising social analysts and philosophers whose work has been published since about 1960. Chapter 1 commences with the version of naturalism put forward in conjunction with the empiricist or, more specifically, the hypothetico-deductive interpretation of natural science. This viewpoint is discussed not because it now commands wide assent but because the most significant anti-naturalist arguments were developed in response to it. Accordingly, most of the first chapter is given over to the examination of a wide variety of anti-naturalist proposals. In the next chapter the more recent formulation of the philosophy of science (often termed the 'post-empiricist' account of science) is introduced. The implications of this philosophy of science for sociology are examined and consideration is given to the ways in which this revised philosophical account accommodates many of the anti-naturalist objections.

The principal advocates of this last position have, however, been philosophers and the remainder of the book is devoted to more detailed investigations of the methodological implications of post-empiricism and of the naturalism debate than have customarily been provided. More elaborate specifications of the purely philosophical issues can conveniently be found in many of the post-empiricist works, for example those by Bhaskar, Keat and Urry, Papineau, and Thomas.[2] For practical purposes the text will makes relatively economical use of notes and references. Readers will be able to pursue points of interest through the range of works cited in the bibliography.

At this point I should make clear that this study is written from the point of view of a sociologist of science. In the first place, this means that the examples of sociological analysis will sometimes come from studies of the scientific community. More significantly, claims about the nature of scientific research drawn from the large and expanding area of social studies of science are introduced in Chapter 3 as a counterweight to the philosophers' accounts of scientific practice which featured in Chapter 2. In particular this entails two critical perspectives on the actual rather than the intended or presumed relationship between philosophical analyses and (natural or social) scientific practice.[3] Firstly, many studies indicate that any procedural rules devised by philosophers require extensive interpretational work before they can be made to apply in specific, concrete situations. Secondly, as many post-empiricist philosophers are willing to concede, the post-empiricist view of science still has large weaknesses and cannot explain at all fully by what stages scientific knowledge has arrived at its present state, nor precisely to which evaluative principles scientists have adhered.

In Chapter 4 these issues are developed to demonstrate that, with regard to the practice of scientific research, post-empiricism contains shortcomings both of prescription and of description. At one level philosophers may be justified in temporarily shelving these shortcomings because post-empiricism seems to be the best philosophical view of science they have. But in the light of these weaknesses philosophers should acknowledge the view of social analysts that philosophy has no indisputable claim to extensive knowledge about the actual operation of science. Chapter 4 also sees the introduction of various internal criticisms of the new philosophy of science and of a series of pragmatic arguments against a scientifically orientated sociology - arguments which were largely neglected during the rise of post-empiricism. When these criticisms are combined with the weaknesses detected in the post-empiricist account by social studies of science the new, naturalistic philosophy of social science begins to look far less robust, particularly as a guide to practical sociological research. The post-empiricist resolution of the naturalist debate can, in crucial respects, be seen to be only an abstract solution to the problems of sociological method. This critical evaluation of the standing of naturalistic social science is applied, in Chapters 5 and 6, to the assessment of a variety of styles of contemporary empirical research. On the basis of these assessments, proposals are put forward for the direction of research activity. These proposals are pragmatic and take both the philosophy and sociology of science into account. They are not offered as formal philosophical proofs. And this is in accord with the overall contention of this book: that if an investigation of naturalism is to yield results for actual sociological practice then the image of science against which sociology is judged must be drawn with all possible attention to the details of scientific procedure.

CHAPTER 1

Empiricist naturalism and the anti-naturalist response

Introduction

There is a widespread assumption in the industrialised world, presupposed for example by many science journalists and teachers of science, that the application of scientific knowledge yields direct material benefits. By contrast, what little success has been experienced in ordering the human world has often been attributed to the personal qualities of politicians and leaders in other spheres, rather than to the implementation of theories of society. Of course this may be unfair to Keynes for instance or to Marxism-Leninism, but as many evils and misdirections can be attributed to social theories as successes. Applied social science does not appear to afford the utility seemingly guaranteed by natural science. And the obvious corollary to draw is that the weakness lies in the way that social theory is generated, and that the route to improvement is through imitation of the natural sciences. Moreover, because philosophers have been concerned with spelling out the characteristic features of natural science ever since the beginning of the 'scientific revolution', their writings have provided a convenient touchstone for would-be reformers of social theory.

Two further kinds of consideration have led sociologists to look towards the methods of natural science, or rather to philosophers' portrayals of these methods. There was, firstly, a growing amount of statistical data about a vast range of human attributes: voting behaviour, income, size of family, leisure activities and so on.[1] This material, generated by and for an increasing number of institutions, contrasts strikingly with the kind of personal knowledge which people customarily have of their social environment and resembles the impersonality usually ascribed to the factual data of natural science. Along with the adoption of particular statistical techniques often first derived for natural scientific use, these factors led the general advance of sociology to be associated with the imitation of science.

Secondly, there was the claim famously put forward by the logical positivists in the third decade of this century that any claim to knowledge which could not, in principle, be subject to empirical test was either true by definition (all whales are mammals) or meaningless (there is one true God).[2] And even though enormous internal difficulties led to the abandonment of this doctrine by the fifties, the premium it placed on strictly empirical knowledge and the exemplary status attached to natural scientific knowledge (as opposed to earlier exemplars such as moral certainty and logic) have persisted. For sociologists there were therefore both the fear that certain cherished notions like progress or social harmony might have no clear empirical referent and thus be muddled if not exactly meaningless, and the certainty that there could be no better model to look to than natural science.

For these kinds of reasons, in the fifties and early sixties many sociologists were actively prepared to look to natural science for indications of how to develop their science. At the same time the empiricist school of the philosophy of science had developed a detailed and sophisticated programme which seemed able to offer precisely the guidance needed by social science. This empiricist philosophy, often known as hypothetico-deductivism, was sufficiently developed that its principal tenets could be encapsulated in a short, compelling book by Hempel while it could also give rise to major, encyclopaedic statements such as that provided by Nagel.[3] Whilst acknowledging the existence of certain internal, technical problems many philosophers were able to proclaim a clear and coherent general viewpoint. Authors like Hempel provided many examples to illustrate the contentions made in their texts and several philosophical authors were, or had been, practising scientists. Accordingly it could readily be maintained that they had successfully formalised the essential structure of science.

Empiricist naturalism

In order to reveal the implications which this view of naturalism appears to hold for the practice of social science it is necessary to present an outline of the hypothetico-deductivist (henceforth H-D) image of natural science. One of the illustrative scenarios supplied by Hempel is of value here. He gives an account of the identification of childbed or puerperal fever.[4] In the middle of the nineteenth century there were at the Vienna General Hospital two maternity divisions and the incidence of the fever was many times greater in one than in the other. Hempel relates how a doctor named Semmelweis was troubled by this fact and set about trying to establish how it could come about. He tried to allow for all possible relevant differences between the patients in these wards: including physical overcrowding, the bodily care and sensitivity of medical treatment accorded to the women, and even the emotional and spiritual influence of the visits of priests to administer the last rites since these visits were more conspicuous in one division than in the other. No attempts to redress these differences yielded any consistent elimination of the discrepancy between incidence rates of the fever.

Hempel records that Semmelweis was running out of ideas on what factors might make the difference between the wards when a colleague of his received a slight wound from a scalpel during an autopsy and subsequently died of sickness closely resembling the fever. Semmelweis concluded that matter on the blade had been responsible for the infection of his colleague. He then noted that in the first division mothers were examined by doctors and medical students who had come directly from dissection sessions whereas in the other wards the examining assistants never undertook autopsies. He concluded that the infection was transmitted from the bodies back to the mothers by the medical staff themselves, and to combat this introduced a policy of disinfection after dissection sessions. Hempel reports that the mortality rate quickly fell below that of the other division.

Independently of the historical correctness or accuracy of this account one can use it to bring out the central features of the H-D view of science. Crucially, the scientist is confronted by a number of facts all of which can be observed and manipulated independently of each other and of the explanation being sought. These observations can be made by virtually anyone and the facts are taken to be unambiguously specifiable. Because of this, correlations can be formed between various sets of facts. These properties of facts and observations constitute the foundation of scientific investigation.

Such investigation proceeds against a background of previously established facts and currently accepted hypotheses and theories. In any particular instance, whether one's interest is led by a practical concern as with Semmelweis or by a more diffuse curiosity, the procedure is fundamentally similar. Hypotheses are generated which link the facts together and then the correlation between these facts can be measured and observed to determine whether the hypotheses actually hold. To make the explanation complete these hypotheses must relate to some general law such as that 'contamination of the bloodstream generally leads to blood poisoning'. Due to their public, incontestable nature facts have primacy and they form the foundation of knowledge because they are open to everyone and because they make up a cumulative stock of information. Hypotheses, by contrast, can be ascribed to particular individuals who propose them; they contribute to the dynamic character of knowledge by establishing new connections between facts and by indicating what kinds of facts are important at any specific time. None the less, hypotheses are not irrevocable in the same way as facts and do not possess the same kind of certainty – it is forever possible that future factual instances will contradict and thus overthrow any hypothesis.

While hypotheses may be related very closely to elaborate theoretical beliefs, as for example the postulation of a new viral strain on the basis of familiar but not precisely identifiable symptoms this is by no means necessarily so. As Hempel makes plain, Semmelweis' hypothesis about the fever was satisfactory even though he had no very clear idea about why autopsies should be connected with re-infection. Apart from the vague notion of blood poisoning, he had no particular theory of why the transfer of material from the dead to the living should be fatal. It is the good fit between the two sets of facts linked by the hypothesis rather than the intelligibility of the nature of the connection which matters most. The idea of dread inspired by the priest or the physical discomfort of overcrowding are readily understood whereas the association of autopsies and infection exists simply as a correlation without any self-evident connection.

This last point relates directly to two fundamental and rather technical aspects of empiricism. Firstly, it reveals the significance of the empiricist analysis of causality. As is well known, as early as the eighteenth century Hume sought to demonstrate that we cannot know anything more about the relation between a cause and its effect than that they are consistently associated. No matter how closely we examine any causal chain it is impossible to 'see' or locate the causation – instead all that there is to go on is the contingent experience of the two things repeatedly happening one after the other. Empiricism takes this point to heart and accepts that all statements about causal connections are really assertions that one kind of occurrence is always associated with another. We are deluding ourselves, or so it is maintained, to think that hypotheses or theories explain why putrid material leads to infection. All our explanations do is record that infection follows contact with such material. Explanations are just like predictions – both are summary statements about sets of events being regularly associated.

Of course the common-sense distinction between things which just happen to be associated (coincidences) and things which have a deeper form of connection can be honoured by saying that coincidences will in the end betray their lack of constancy. But observation or inspection of particular instances of association does not allow you to see whether a cause is present or not. Once you accept that hypotheses and theories do not penetrate to a mysterious entity – causes – it is easy to regard them as instruments or devices for recording the correlation between facts. Empiricism accordingly tends to favour an 'instrumentalist' view of theories and hypotheses: they are means for the calculation of the world's future factual states and are valued only in so far as they serve this function well.

The other fundamental point also relates to instrumentalism but deals principally with the structure of theories. Theories, or statements of connections at a high level of generality, are assumed to make up rigorously deductive hierarchies, so that lower level statements can be deduced from (combinations of) higher ones. As an example, the statements concerning the revolution of the planets around the sun can be deduced from the theory of universal gravitation and the statements describing the fundamental law of motion. From the lowest levels of these theoretical structures deductions can be made about single instances, and these in their turn can be checked against the world of facts through the use of 'bridge principles'. Hempel illustrates these bridge principles or hypotheses with the instance of gas pressure being taken as corresponding to the impact of individual molecules on the containers in which gases are held. Such an assumption is indispensable in moving from theoretical statements to observable properties; the correspondence between fact and theory is handled by means of such bridges.

Accordingly instrumentalism is important again here because the theoretical conceptual structure can be regarded as a fiction or device for interpreting the world of observed instances which is 'clamped onto' the world of perception through bridge hypotheses. Since these bridges are virtually definitional, theories can be abstract structures with only a formal correspondence to the world. Both with conceptually remote items such as sub-atomic particles and with other remote, inaccessible concepts like the Big Bang or the origination of the universe such a view is thoroughly plausible. There is nothing more that could be demanded of the notion of mesons or quarks or the creation of matter than that the implications of these concepts cohere with what is known through observation. Such theoretical entities will forever stay on the other side of the bridge principles so that the only conceivable criterion for their acceptability is that they correspond with what appears on this side of the bridge.

This means that the prime test of a theory is that deductions from it should cohere with the growing set of facts known about the relevant object, and this coherence can be established through the theory's success in giving predictions and explaining past instances. The scope of a theory is generally increased by elaborating on its details – say in Semmelweis' case by saying more about what putrid matter should mean. But a theory can be good even if it is of the most sparse instrumental sort (as was Semmelweis' initially). Theories do not need to be elaborately understandable (like the notion of dread of death inspired by a priest) to be formally acceptable.

The object of scientific investigation can therefore be seen to be the generation of increasingly many accurate laws specifying the nature of connections between observable facts or between hypotheses which in turn relate to such facts. These laws are built up using intellectual resources generally available in the existing stock of knowledge and it is desirable that laws should be co-ordinated into the hierarchical conceptual structure.

In some ways this may appear to be a bleak formal depiction of science, but it is important to realise that H-D empiricism is aimed at disclosing the logical framework behind any common-sense notions of scientific activity. H-D empiricism concerns the essence of science since it aims at demonstrating what logical status scientific pronouncements have. On this view science can be characterised as the rigorous and rational pursuit of the attitude to the world implied by these facts of logic. Hempel admits that people have always sought to understand their environment – hence the existence of myths and anthropomorphic cosmic theories – but such an urge does not adequately characterise what lies behind science:[5]

> Accounts of this [mythic] kind undeniably may give the questioner a sense of having attained some understanding; they may resolve his perplexity and in this sense 'answer' his question. But however satisfactory these answers may be psychologically, they are not adquate for the purposes of science, which, after all, is concerned to develop a conception of the world that has a clear, logical bearing on our experience and is thus capable of objective test.

The philosophical analyses of the H-D empiricists aim to reveal what logical features of science distinguish it from these mythic kinds of explanation. The first feature is that the explanations given must be susceptible of some form of empirical test. Unlike explanations in terms of, say, demonic powers which are inherently wilful and inscrutable so that they can never be specified in advance, it is essential that scientific explanations can be tied down in the physical world so that conditions for an empirical test can be set. The other feature is called by Hempel 'explanatory relevance' and refers to the nature of the items identified as cause and effect. The connection being made must be licensed in some way, either through precedent or through empirical demonstration. In this sense, witchcraft is not explanatorily relevant to the inception of illness while the transfer of organic matter from a dead body may be.

Without for the moment going into any of the criticisms it may be possible to make of these depictions of natural science, it is important to examine the implications their acceptance was claimed to have for the direction of social science. Clearly, much of the logical positivists' insistence that the only viable factual knowledge is scientific endures in Hempel's treatment of explanation. It seems that any satisfactory explanatory undertaking must conform to the pattern set by natural science and that H-D empiricism specifies how science must be structured. This sentiment receives explicit statement in van Dyke's book on political science written at the height of H-D empiricism's influence:[6]

> *science* and *scientific*, then, are words that relate to only one kind of knowledge, i.e., to knowledge of what is observable, and not to any other kinds of knowledge that may exist. They do not relate to alleged knowledge of the normative - knowledge of what ought to be.

While allowing in principle that there may be other kinds of knowledge than the scientific (perhaps religious knowledge or political wisdom) their actual status is rather undermined by the use of the epithet 'alleged'. For van Dyke the evident path towards success for political theory is towards a scientific dependence on facts and away from the normative and moral tenor of traditional theorising.

Thus the first requirement of H-D naturalism for social science is that it should be scrupulously factual and not evaluative . Only in this way can a scientific methodology be adopted. And there are also more concrete methodological directives. Definitions of social phenomena must be made as precise and as objectively free from common-sense meanings as possible. Natural science has progressed by releasing its concepts of motion, of what things are composed of and so on from the dominance of everyday and mythic conceptions, and social science must do the same through the adoption of criteria of empirical testing and explanatory relevance. Similarly, neither description nor explanatory laws should be formulated in terms relating to local or specific phenomena. Instead they should be designed to achieve maximal universality in the same way as physics. Correlations between facts observed by analysts should accordingly be made over as wide a range of instances as possible so that consistent, 'causal' associations can be separated from chance concatenations.

In connection with this last aim, statistical methods are particularly welcomed since they appear to provide a convenient and objectively impersonal means for establishing the degree of support for hypotheses. Given that in many circumstances researchers are liable to unsystematic human interaction with the objects of their investigation, the opportunity offered by statistics to deal with large numbers of cases on a rigorously standardised basis appeared particularly attractive. Allied to the statistical concern with correlations is the tendency to accept an instrumentalist orientation to theorising. Just as statistics can reveal societal facts without saying why correlations occur, so also can theories allow the prediction of social trends without dwelling on any underlying causal process. Such a view of theorising is perhaps best known within economics where models of the economy with predictive capacity but little internal elaboration are happily tolerated.[7]

In sociology and psychology it is possible to see this instrumentalist approach tied closely to a desire to move away from untestable concepts. It can readily be suggested that people's wills or thoughts and desires are every bit as untestable as the malevolence of witches: such personal attributes are invoked to make sense of human actions but there is no check on those internal states which supposedly underlie observable behaviour. Accordingly, the drive towards testability inclined social scientists to concentrate on behaviour and not internal states, and although behaviourist explanations do not appear as complete in some intuitive sense as reasons based on desires and will, this consideration does not matter if theories are regarded as mere instruments. In this way the directives of H-D naturalism readily appeared to offer a coherent programme for the social sciences.

Of course these directives are very general and can be implemented in many different ways as the variety of examples mentioned even in this short section reveals. In addition, the philosophers of H-D naturalism were not entirely in agreement on all matters, and taken more generally there have been many versions of positivism adopted more or less literally by different social scientists.[8] For example, as is demonstrated by Keat and Urry, Durkheim espoused particularly strict principles concerning the inaccessibility of unobservable personal states, whilst actually alluding to such states in his formulation of the suicide problem.[9]

The overall trend of the directive is, however, very clear. Theory and observation are distinct epistemological categories and should be well separated. Any hypothesis or theory should be open to direct empirical test. Values must be excluded from the stock of knowledge, and only observable facts and indices of facts should be attended to. Resulting from this would be a cumulative stock of factual knowledge and a growing body of theory, improving and refining common sense and affording an increasing predictive power and associated capacity to control the social world.

Anti-naturalist responses

H-D empiricism can be seen to provide epistemological reasons why social science should model itself on natural science since it argues that there is no other route to sure knowledge of any empirical matter. But there is a converse argument built into this contention. In so far as do anyway know things empirically about other humans and about the institutions which populate our societies then we do possess another model for sociological knowledge. It remains to be seen whether this latter model represents a proto-scientific form of the earlier one or whether it constitutes a real alternative.

Anti-naturalist responses to H-D naturalism comprised anything but a uniform front in terms of the arguments deployed and the backgrounds represented, the latter ranging from continental Marxists, to American fieldworkers, to Oxford dons reflecting on linguistic proprieties. Their attacks were seldom mutually supportive and were aimed at different facets of the connection between H-D empiricism and social science. Yet in one sense they were united in arguing from material furnished by the model for sociological knowledge afforded by our everyday experience. This is why the points they made can appropriately be discussed together, arranged in terms of issues rather than chronology or cultural origins.

Few of the anti-naturalists were, however, philosophers of science. Thus to anticipate myself a little, while 'post-empiricism' was developing and as it were internally transforming the philosophy of natural science, there appeared in relative isolation from this process a series of objections to local tenets of H-D naturalism. Post-empiricism now claims to be robust enough to accommodate and resolve all such objections, thereby saving the naturalist spirit of H-D empiricism. Its proponents argue that these objections do not militate against naturalism *per se* but only against some of the peculiarities engendered by H-D empiricism. But rather than move directly to post-empiricism it will be beneficial to consider the range of anti-naturalist objections for a number of reasons. Firstly this descriptive approach will help the reader to situate the various methodological studies they may come across in general sociological reading. Secondly it will clarify the implications which post-empiricism holds for social scientific practice, and finally it will indicate that post-empiricism does not embrace all the possibly valid objections to naturalism. These outstanding objections, together with the weaknesses detected in post-empiricism itself, will form the subject matter for the substantive methodological analyses of Chapters 4 and 5.

Pragmatic objections to a naturalistic sociology

The feature common to these pragmatic objections is the claim that, however accurate epistemologists' descriptions of natural science, there are contingent peculiarities of social knowledge which make these philosophical depictions inappropriate. For example, it may be said that social science is concerned with unique events or instances (say with Christianity as a particular religion) and therefore does not have the generalising interest which the H-D view of science presents as universal. However, that this is an artificial basis for a distinction – albeit one encouraged by the examples taken by Hempel – can be readily seen from the natural scientific study of singular occurrences. Comets may pass the earth once and collide before returning, or one may want to explain why some feature of landscape, say the shape of the Wash, happens to be the way it is. Two considerations arise from these comparisons: firstly, any one thing is not totally unique. Any particular comet is in some respects like other comets, and comets in general are in striking ways like planets, stars and so on. If comets, or anything for that matter, were radically distinct they could not be described in the terms used to depict other things and thus could not be incorporated into any general theory. Secondly, the shape of the Wash – just like the inception of the illness of Semmelweis' colleague – is explained as the entailment of more general relationships and tendencies which we know about. The uniqueness of some occurrence does not mean that general principles are inapplicable to it.

Another interpretation of uniqueness is in terms of the historical singularity and unrepeatable nature of social occurrences. The human world is never static and its future condition is open since the environment created by today's activity, which will provide the conditions for future action, is always novel. Because of historical change the social world might be thought to lack the uniformity of atoms and forces on which physical science is built. However, natural sciences like geology are also historically singular. The moving continents create ever new conformations of the land and sea with novel climatic consequences, and the gradual cooling of the earth means that environments are never truly stable. Yet geology is not deemed less of a science on this account. It might be suggested that geology is one of a few sciences (including meteorology) which are dependent on the laws and findings of other sciences for their ultimate explanatory principles, but even if one were to admit this much for sociology it would not count against the pursuit of naturalism at all.[10]

One might also argue that sociology should not be modelled on natural science because it does not depend on experimentation. However, if this claim is taken literally it is not entirely true since there have been experiments in the sociology of small groups.[11] If the claim is not taken this strongly it appears to be as true of certain facets of geology as of sociology. Here again the analogy with geology, whose practitioners make observations in the field as do social analysts, shows that no absolute distinction separates sociology from science. A related case for anti-naturalism can be made on the grounds that sociology does not make predictions, but again this is not strictly correct. And in so far as it is correct - i.e. because social science does not make many good predictions - it is also correct of subjects like evolutionary theory whose scientific standing is not attacked. This issue of whether sciences must make predictions will be raised again in the next chapter as part of the discussion of post-empiricist views on explanation and instrumentalism.

It should be recalled, however, that the instrumentalist view of theory is very specifically organised around predictions. And it is important that this capacity for prediction is not ruled out in principle, even though it may often be thwarted in fact, by self-fulfilling prophecies. These are instances where knowledge of a prediction enters into people's estimation of a situation, changes their attitude and thus may be responsible either for the success or the failure of the prediction. This might occur in an auction, say, where a prediction that people are willing to bid high may actually influence their propensity to offer large sums or discourage potential bidders and therefore lower the price. But in these cases it is leaked information affecting the antecedent conditions and thus altering the whole situation which 'spoils' the prediction. The actual situation is no longer the one for which the prediction was prepared. Had the sociologists kept the knowledge to themselves the prediction would have stood a better chance of holding. Thus it is not the predictive weakness of sociology but the capacity of the subject of study (i.e. people) to act on the basis of knowledge about themselves which sets social science apart. This issue relating to action and free will is handled in a later section.

Sometimes it is claimed that experience is essential to knowledge of society in a way that is not required in natural science. Clearly, however, certain kinds of experience are required in natural science: beginners often cannot get experiments to work, and at the forefront of knowledge experience is important at least in framing hypotheses as was the case with Semmelweis. If this is the extreme of the weak sense in which experience is vital to social knowledge, there is also an extreme strong interpretation. This demands that in order to understand someone else's position one has to replicate their experience exactly. But given the unique biography of every person this requirement would lead all communication between people to be logically flawed - this is too strong a formulation of the place of experience in social knowledge. Thus there appears to be no absolute divide between the type of experience necessary for natural scientific knowledge and the kind of experience on which social knowledge depends. Admittedly, in the case of exotic or unusual experiences it does seem that the analyst faces a severe problem since, for example, an investigator can have no conception of a social practice which depends on taking peyote unless the investigator joins in the consumption of the hallucinogen. However, this argument is significantly related to the point mentioned above about the uniqueness of occurrences in the social world: there is no experience which is truly unique. In certain respects the effects of peyote are like the effects of alcohol and peyote-inspired hallucinations are like exceptionally vivid dreams. Because everyone has experienced some range of emotions and somatic reactions, people can derive an understanding of each other's experiences through various, often metaphoric, descriptions and through seeing the effect that an experience has on others. Thus through, for example, prisoners' descriptions and through studying, say, the physical deterioration of long-term prisoners one can derive an understanding of what it is like to serve a life sentence without suffering the experience. Clearly, the testimony of people who have undergone the experience will be vital but there is no proof that only they understand the phenomenon. Considerations from social knowledge again fail to provide a firm demarcation between sociology and natural sciemce.

There is another argument against naturalism which relates to experience. It is suggested that without some experience of a culture's customs the investigator would not even understand what certain actions were about, let alone comprehend what it was like to experience those actions. Without, for instance, some experience of leisured, Western life analysts would be mystified by clay-pigeon shooting. The lack of experience would not only debar them from perceiving the satisfactions of this sport but would prevent them from seeing it as a sport at all, as distinct from, say, a rite or ceremony. But fundamentally this argument is not about experience, rather 'experience' is being employed as an expression for the meaning which some action has within a culture. That 'experience' is being used in an unusual sense can be seen from the fact that detailed information from a native on the meaning of clay-pigeon events would serve to remedy the investigator's inexperience. Consequently this argument does not concern personal experience, rather it deals with culturally specific meanings. The problems generated by these meanings will be considered in a later section.

Lastly, some might argue that in society people know best: again this is ambiguous between a liberal political sentiment that one knows one's own best interests (a sentiment which is ill-founded in a society where people's lives are governed by laws and processes which they do not fully understand) and two other claims. It may be that social situations are so complex that only people involved in them have the time, opportunity and motivation really to get to the bottom of them. Alternatively, it may mean that people's activities are partly constituted by their own understanding of what it is they are doing. For example, two people attending a church service may not be doing the same thing if one regards attendance as a guarantor of salvation while the other sees it as an opportunity to find some personal peace. The first of these points is actually another formulation of the question of the complexity and uniqueness of social occurrences; it thus forms no decisive argument against naturalism. The latter represents another facet of the problem of social meanings to be examined below.

Thus, of all the pragmatic justifications for anti-naturalism only those which form a sub-set of other, broader objections (like the problem of meanings or free will) and the point about the understanding of exotic experiences pose any threat to naturalism. All the other issues, such as historical uniqueness and complexity, merely amount to difficulties in implementing the naturalist's programme. They point towards no alternative route to knowledge. For example, the potentially most intractable of these problems – the possibility that the social world may change so rapidly and so complicatedly that it actually outpaces the capacity to update knowledge of the world[12] – only shows that naturalism may be impracticable. It does not show that naturalists are wrong to aim for knowledge on a scientific model, nor that there exists a better exemplar for knowledge of the social world.

Logical objections to a naturalistic sociology

The set of objections which can be termed logical are also founded on comparisons between commonly available social knowledge and the empiricist depiction of scientific knowledge. However, in these cases the contention is not simply that the empiricist model does not do justice to the peculiarities of social knowledge, but that the kind of knowledge involved has radically different logical properties. The first argument in this class concerns the possibility of linguistic comprehension. Clearly, much social scientific work involves evidence obtained from what people say or write: interview schedules and Marx's analysis of political economy both fall in this category. One view of the procedure employed in this use of language presents understanding as the recovery of what people intended in saying what they did. Thus, if a chemist says 'Keep the container at thirty degrees', what the utterance means is given by what the speaker intended the expression to mean: in this case, that the chemist desired the bath to be maintained at a constant temperature for the purposes of some experiment.

On this view, utterances are fundamentally a sub-species of action since their essential features are tied directly to the intentions they serve. This connection between language and action is spelled out explicitly in Searle's analysis of what he terms 'speech acts'.[13] He examines the meaning of utterances in terms of the action accomplished through the utterance (for example, expressing gratitude). The properties of speech acts and their relation to naturalism will be considered below along with actions more generally. But returning to the claim that linguistic comprehension amounts to the recovery of the intention behind utterances, there are some evident difficulties in accepting this as a general account of meaning and understanding. In the same fictional laboratory as that in which the chemist spoke, one can imagine a novice technician who has been told to heat a sealed container of alcohol to 30 °C. Instead of applying heat gradually by immersing the container in the chemist's water bath, the technician decides to complete the job hurriedly using a bunsen burner. A senior colleague noticing this might well say 'Go ahead and blow up the lab' as a way of ridiculing some sense into the technician. Now, exactly the same words could be employed to initiate the destruction of a biological laboratory by a group protesting against the experiments carried on there. This example indicates that identical strings of linguistic units can be employed in differently-intending utterances, and of course different words can be substituted whilst expressing the same intention: the colleague could have said 'That's it – destroy the building'.

The intentionalist account of meaning is therefore inadequate since it does not explain how intentions and linguistic tokens (whether words or sentences) can both be related and yet relatively independent. The words 'blow up the lab' must have some degree of meaning of themselves, independent of any intended use of them, so that particular utterances are, as it were, applications of this meaning. The intentionalist meaning of utterances can be said to be parasitic on the meanings of the tokens employed. Accordingly, one is faced with a level of what might be termed conventional meanings which occupy a different dimension from actors' meanings *qua* intentions. And any social science or social knowledge which involves understanding language is dependent on this set of conventions. Furthermore, knowledge of this convention does not consist in an explanatory inquiry into how it came about; instead it appears to be an entirely non-naturalistic acquaintance or familiarity with the code. Unlike the intentionalist view of meanings which directs attention to why the utterance was made (and thus has at least some naturalist features) conventionalised meanings appear to be completely symbolic. Because of sociology's dependence on this kind of linguistic comprehension it is tempting to conclude that sociology must be radically distinct from natural science.

It is however important to determine precisely what this argument serves to establish. It does demonstrate that there is a form of knowledge (or competence as Chomsky would label it[14]) which is foreign to the H-D empiricist depiction of valid knowledge; for example, that trees are called 'trees' and not 'trousers' is hardly susceptible of useful empirical investigation – every noun could just as easily be renamed as something else. It seems natural to move on from here to say that while social science essentially depends on this kind of knowledge, physical science does not. It appears necessary to the gaining of social scientific data whereas natural scientific experiments do not speak.

Against this demarcation one must insist that while natural scientists' experiments do not speak, the scientists themselves do. They gain their knowledge and training through conversation and reading and are thoroughly dependent on linguistic competence – the scientific community produces knowledge corporately and the success of this community depends on the transparency of linguistic comprehension. At this level the social scientist is in an exactly analogous position. And if it is a necessary postulate of sociological investigation that communication between social scientists is sufficiently transparent for them to convey information and arguments to each other, then it can hardly be assumed to be an obstacle to investigation that sociologists have to converse with their subjects. Of course, subjects may not want to be known about or they may have beliefs and motivations which the analyst regards as very curious, but any problems arising here are to do with the intentionalist meanings and not the linguistic meanings of what the subjects may say. Such problems relate to the meaning of actions and speech acts, and will be considered below. Thus the argument from linguistic comprehension shows that there is a distinct form of knowledge but it appears that this knowledge is actually presupposed by both social and natural science so that it cannot discriminate between them. Unless social science were to restrict itself to linguistic analysis (interpreted very strictly to exclude even speech acts) then it requires knowledge distinct from linguistic competence, and more akin to natural scientific knowledge.

The argument from linguistic comprehension can be strengthened by claiming that other significant facets of human culture comprise symbol systems and thus require a related form of understanding. Fundamentally this is the claim of structuralism and some versions of semiology.[15] For example, it can be contended that many facets of culture such as standards of proper dress, dietary customs and greeting procedures have no intrinsic rationale: there is no good reason why academics should wear mortar boards rather than some other hat. Nevertheless, these details of behaviour may be very significant to members of the culture despite the fact that they seem as arbitrary as the fact that we call trees 'trees'. This similarity to language suggests that exactly the same kind of understanding may be involved.

However, further investigation indicates that this is not the case. Whilst langauge can still continue to function even if its arbitrariness is recognised this hardly seems true of other cultural systems. For example, rules of kinship which specify which marriages are forbidden because of incest tend to degenerate if the arbitrariness of the kinship terms (say with respect to actual hereditary connections) is revealed. This suggests that other symbol systems no matter how arbitrary they seem to us, are taken to be literally veracious (anthropologists indicate that this is even true of many cultures' conceptions of the relationship between words and the things they name). Moreover, the invocation of these cultural systems has aspects which resemble the use of language in speech acts, as for example snubbing one's host by dressing inappropriately. Such uses of these conventions share the properties of intentionalist meanings and actions more generally; they cannot therefore be comprehended on the basis of knowledge analogous to linguistic comprehension alone.

The status of actions themselves provides the next logical objection to the assimilation of sociology to the natural sciences. As was seen above, in the H-D view of the natural sciences a result can be taken to be properly associated with a cause only if the events concerned are independent. Someone's being a bachelor cannot be the cause of their being unmarried, nor can something's being a whale be the cause of it's being a mammal, because in both cases the terms are logically or internally connected. The events or states described in these instances are not logically independent. On the other hand, discrete billiard balls are independent entities and a collision between them can be seem as the causal episode which resulted in a transfer of motion between them. This logical objection centres on the suggestion that human actions do not meet the conditions imposed by empiricism on the identification of causes, and that in important ways human actions are understood through internal connections.

In the first place it is rather difficult to come up with unambiguous identifications of actions. It is not possible to make the convenient move of converting descriptions of actions into accounts of behaviour. The same behaviour, such as the drumming of one's fingers, could constitute fidgeting, playing a piano, or tuning a piano or even annoying a sensitive musician. Similarly actions, such as greeting someone, can be accomplished through a variety of behaviours - nodding, waving, speaking and so on. Worse still, an action may be achieved through not behaving at all as in failing to stand for the National Anthem. Of course this is not of itself to show that actions cannot be the object of naturalistic analysis, but it does remove the obvious means of incorporating actions into the world of natural objects.

There are two principal arguments for the non-naturalistic character of our understanding of human action. The first is based on the claim that actions are usually rendered intelligible in terms of the reasons which lay behind or supported some action. Initially it may be tempting to view these reasons as the putative cause of the action but, the argument runs, this is to overlook the point that reasons are evaluatively rather than causally assessed. The inquiry is not whether one's reasons were in fact one's reasons but whether they were good enough reasons. If a teacher excuses some misbehaviour and gives as a reason an opposition to the punishment of children, one does not doubt that this was the reason for excusing the child though one may doubt whether it is a good reason. Thus the contention is that reasons are assessed on a dimension which is alien to natural science, and that the dimension of interest to natural science is typically beside the point in matters of social knowledge.

The seond objection again relates to the status of reasons. It is suggested that reasons are not separable from actions in the same way that causes are (and must be) separable from effects. For example, if someone is asked why they are eating a choc-ice they may answer 'Because I wanted an ice-cream'. Although this reply is shaped like a causal explanation the reason and the outcome are more closely linked than is the case with physical causation. For this reason, Taylor argues,[16]

> the two are not contingently connected in the normal way. We are not explaining the behaviour by the 'law', other things being equal, intending X is followed by doing X, for this is part of what we mean by 'intending X,' that, in the absence of interfering factors, it is followed by doing X. I could not be said to intend X if, even with no obstacles or other countervailing factors, I still didn't do it. Thus my intention is not a causal antecedent of my behaviour.

The reason is tied to the subsequent action through a logical and conceptual relationship rather than through sheer factual co-occurrence. Thus reasons relate to actions in something like the same way as bachelorhood to unmarriedness and social knowledge which depends on understanding this connection cannot straightforwardly be assimilated to naturalistic science.

These points about action are often connected to the discussion of rules since it can be contended that many forms of action are rule-governed. Clearly this is not true to the same degree of all rules: playing football is (usually) more rule-constrained than an expeditionary trip to the Arctic. But it can be asserted that social activity is hedged around with rules, especially if 'rules' is interpreted in a broad sense. With regard to sociology, rules have probably attracted more empirical attention than actions since rules are inevitably general phenomena within any culture and therefore offer opportunities for repeated observation which compare favourably with the uniqueness and individuality of actions. However, despite the extensive references to norms and rules in much naturalistically-inclined social research, philosophical analysts have drawn attention to anomalies which arise if social rules are taken as resembling natural laws - the apparent rules of nature. The idea of viewing social rules as analogous to natural laws is encouraged by H-D naturalism since laws and generalities are highly valued, but there is an obvious dissimilarity. Non-conformity to a natural law either indicates that the law has been wrongly formulated or that the observation was erroneously made. These two possibilities also exist for social rules, but there is the further possibility of purposeful disobedience. Such a rule may be overwhelmingly disobeyed and yet can properly be said still to exist, if for example the powers of enforcement are temporarily depleted because of a police officers' strike.

In general, rules have two properties which are unlike natural laws: firstly, they are orientated to consciously by the actors whose behaviour is said to be constrained by the rules. Rules do not so much express regularities about people's actions as regularities to which people aspire in acting. Secondly, rule-related activity is susceptible of evaluative assessment in a similar way to the reasons people give for acting. It is possible to say that someone misapplied a rule or that their action was not justified by some rule which they cited. Societies appear to abound with rules and in so far as rules, just like reasons, can and do sustain types of question which are foreign to naturalistic knowledge there are grounds for resisting naturalism in sociology.

A further dimension on which social action appears to diverge from natural occurrence concerns the intentionality of social action. Returning to the intentionalist theory of meaning it can be seen that the chemists' intention in speaking about the water bath is not the same thing as their reasons for saying what they did. Therefore, even if it were granted that reasons are rather like causes so that they could be understood naturalistically there is still this intentionalist facet of human activity. Roughly speaking, intentions are people's own understanding of the undertaking which they are involved in, for example 'worshipping God', 'repairing the car' or 'preparing an experiment'. Whatever the reasons for these projects - and whether or not these reasons are subconscious and thus hidden from the actors - actors can themselves customarily provide a description of what they are doing. And since an understanding of each other's activities usually depends on a grasp of intentions rather than an exhaustive examination of their reasons or motives, and because there is no correlate of intentions in the objects of natural science, there appears to be a different emphasis in social knowledge from the kind of knowledge a naturalistic social science could offer.

The final logical doubt about the naturalistic standing of sociology derives from virtually the opposite methodological direction; it relates to the reliance of many kinds of social scientific research on statistical results. The argument is that, whereas natural science deals with constant conjunctions and invariable causal sequences these never appear in social scientific findings, which concern probabilities and statistical correlations. Thus the query runs: if, for example, sixty per cent of the children of fathers of socio-economic class one themselves obtain occupations of the same category, what does this correlation or conjunction establish? It can then be suggested that this result is merely the summation of unique, individual decisions and life-opportunities developed by particular persons. Of course certain areas of natural science are also founded on statistical associations but in these cases this is because the underlying variables are assumed to have a random component. Surely, it is argued, statistical associations occur in sociology not because people are random, but because they are individual: each is decisive and explicable in their own right but not as an instance of a generality. Therefore individualists deny the value of accepting generalising natural science as the exemplar for sociology - personally specific knowledge, familiar from everyday social knowledge, is the preferred model.

Transcendental objections to a naturalistic sociology

The arguments of the last section were largely raised by analysts concerned with the philosophy of action and centred on putative differences between scientific knowledge and the kind of understanding customarily applied to human activity. The propositions to be examined in this section have been generated by people concerned with sociological knowledge in a wider sense than the understanding of particular actions. Their interest is in factors which govern the place of understanding in all possible human societies, that is factors which transcend particular historical or personal considerations. Accordingly the arguments are of a more schematic and programmatic nature.

The most truly transcendental claims are those proposed by Habermas and, subsequently, Apel.[17] As a prelude to his philosophy of social science Habermas sets out to characterise natural scientific knowledge. He adopts in large measure an instrumentalist view of theoretical notions, although it is not entirely clear whether he takes instrumentalism as a minimum condition for, or the ultimate condition of, knowledge.

That is he agrees that, say, mesons must be justified on an instrumental basis, but he does not specify whether they might not also be a real constituent of the world. However, he introduces a crucial twist to instrumentalism. He denies that knowledge has a single, direct orientation or correspondence to the object of knowledge; an analogy from the study of perception can help to make the point here. Our visual apparatus is supplied with an enormous amount of information all the time that our eyes are open yet we actually register only a minute amount of that detail. What we see is intimately related to our purpose in looking so that, for example, while we may often have cause to look at banknotes we seldom see the watermark. And at a more fundamental level, without some organising and selecting our visual apparatus would not register chairs, tables and so on at all but rather a changing field of coloured patches. The perceptual world is potentially so rich in detail that without some discrimination and restriction we would be overwhelmed by particular details.

In the same way, Habermas suggests, all manner of things could be perceived and known about the world so that in order for our knowledge to be manageable it has to be organised in a selective, restricted way. He argues that scientific knowledge selects and organises its orientation to knowledge around the possibility of instrumental control over natural objects. So, in his sense, both the theories and the observations of natural science are instrumental. For Habermas, this instrumental attitude corresponds to and is explained by an interest of the human species in manipulating and controlling the world's objects for the sake of survival. Thus, from the point of view of the species, the interest appears to be transcendental; since this is so only from the human viewpoint Habermas describes the interests as quasi-transcendental. The constituents of the world and the properties they exhibit do not reside entirely in the world but are picked out and sustained by human activities – a different race of beings could just as legitimately comprehend the world in an entirely alternative fashion. The connection between human undertakings and our conceptual system can be illustrated through the notion of hardness. Habermas suggests that the property of hardness does not reside in diamonds and so on, but is an expression of how certain materials respond to human action.

Habermas' objection to naturalism in social science is developed through the introduction of two other human interests: one concerned with increasing mutual comprehension – which directs the literary, communicational and historical disciplines – where the meaning of concepts is organised in conjunction with the aim of enhanced understanding. Secondly, there is a series of disciplines, including sociology, psychoanalysis and certain aspects of philosophy, whose goal is to increase human self-understanding and self-determination by revealing and helping to overthrow limitations on thought and action. These disciplines are said to have an emancipatory interest. Accordingly the error of H-D naturalism can be diagnosed as the illegitimate modelling of all knowledge on the instrumental interest. Clearly it is possible to conceive of humans instrumentally, either as physiological entities or as, say, value-less slaves who are counted amongst the livestock. Indeed, it is easy to view behaviourist psychology as expressing just this instrumentalist orientation. But for Habermas considerations of reason dictate that an emancipatory rather than an instrumental attitude should be adopted towards the human sciences.[18] Thus he appears to have a systematic basis for differentiating between natural science and sociology, and for rejecting a naturalistic exemplar for social knowledge.

However, when it comes to specifying the nature of this emancipatory body of knowledge Habermas' proposal is less satisfactory. He maintains that, like psychoanalysis, critical sciences require orientations drawn from both the other interests. In the psychoanalytic encounter, it is suggested, the initial interpretative work of formulating the dream and so on relies on communicational skills. The interest is directed towards adequate comprehension. Then the analyst applies the instrumentally-orientated theory concerning the causes of neurosis or mental disturbance to the patient's account. Finally, the cure is not effected by a scientific diagnosis and application of treatment (as on a medical, somatic model). Instead the patient is led to an interpretative awareness of the conditions behind the complaint. In the light of this awareness self-improvement can take place. Such therapy contrasts starkly to a purely instrumental lobotomy, for example, where the whole episode is distanced from the patient's interpretative facilities and self-determination. But even though psychoanalytic treatment is shown to differ from surgical treatment, it is not at all clear that Habermas' account of the essence of the psychoanalytic method furthers our understanding of this difference. He does not show how the instrumental and communicational interests interact in the service of the emancipatory interest. Neither is it clear, because of the involvement of instrumental reason in psychoanalysis, whether it is instrumentality itself or simply its improper application that is to be excluded from sociology. The apparent rigour of Habermas' transcendental schema evaporates when one examines the implications his anti-naturalism should have for sociological practice.

The other author to develop a transcendental interpretation of the distinctness of social knowledge is Winch; his contention can be viewed as the ultimate extension of the pragmatic objection, mentioned earlier, that people know best.[19] He claims that what is peculiar to persons, as opposed to material objects, is that their actions and the state they are in are fundamentally constituted by their own understanding of their situation and by the meaning their activity has, as identified within their culture. Here, it seems, is an attempt to specify precisely the kind of differences Habermas was cataloguing in terms of interests. Winch's argument can also be spelled out in a methodological fashion: in order to identify what someone's activities amount to one has to know what those activities are said to comprise in the relevant culture and whether the person is acting with respect to those meanings. For example, it is only possible to know that someone is doing a rain dance if we know that they believe that that is what they are doing and that rain dances are recognised within their culture. Whereupon, since their activity is constituted by these specific meanings, it is logically nonsense to deny that rain dancing is what the person is doing. It is necessary to accept the person's interpretation in order to grasp what they are doing and, Winch argues, once this acceptance is made it cannot be reversed without surrendering our grasp of the action's significance.

According to Winch, radical conclusions follow from these considerations. For one thing, even when one has entered into the world of meanings which characterise an alien culture or sub-culture and begun to understand members' activities on their terms, it is not possible to engage in the analyses recommended by naturalistic social science. The use of explanatory notions foreign to the culture, necessary for making cross-cultural comparisons or for explaining cultural institutions in terms compatible with empiricist notions of causality, will inevitably involve a perversion of the culture's meanings and thus a denial of the premises of acceptance which allowed entry into the culture in the first place. At most, analysts can use concepts which have a direct translation into terms available in the culture – for instance, Winch suggests that some academic economists' notion such as liquidity preference could be used in a study of Western business circles since[20]

> it is logically tied to concepts which do enter into business activity, for its use by the economist presupposes his understanding of what it is to conduct a business.

The methodological core of Winch's position is aptly summarised by his critic MacIntyre:[21]

> sociological knowledge is the kind of knowledge possessed in implicit and partial form by members of a society rendered explicit and complete.

People's knowledge cannot be bettered since that knowledge actually makes up their society – the analyst can only hope to make the culture comprehensible, and for that purpose alone an analytical account may be superior. This is, Winch maintains, quite unlike the position of the natural scientist.

Indeed, from this perspective, sociology is more akin to philosophy than to natural science since it is not aimed at recording regularities displayed by logically independent phenomena; instead it is concerned with the elaboration of conceptual systems. Because actions are constituted by the understandings of actors, actions relate to each other less like causes and effects than like ideas which entail one another. In general support of his position, Winch adduces arguments about action similar to those discussed in the section on logical objections. He stresses that people's reasons are commonly assessed for justifiability rather than for causal efficacy, and he emphasises the importance of rules in making up the fabric of a culture. Since all languages contain a negative operator and since rules are formulated within language, the possibility of non-compliance with every rule is built into all cultures, whereas not every scientific law could potentially be contravened. Winch ties these issues together by according language and linguistically-mediated meanings the central position in every culture. Each culture consists of a closed, or 'hermetic', world of meanings - often termed a 'language-game' - and is both sustained and to be understood *only* through this language-game. Even the notions which lead us to think that cross-cultural comparisons can be made (for example the idea that either our Western beliefs or traditional African beliefs must be false) are themselves notions which are only available to us through language. To make the comparison on our terms is merely to assume that our beliefs are actually superior; there neither is nor can there be a neutral medium for comparing different cultures' beliefs. Thus while Winch does not deny the actual validity of natural scientific knowledge nor assert explicitly that all knowledge is as good or bad as any other, he does insist that there are no absolute grounds for assessing beliefs. All standards are internal to some culture or other and appropriate only to that culture.

The concept of language-game helps to clarify what status Winch regards sociological understanding as having: the only sense in which one can understand any game is to know how to play or undertake it according to the rules. To understand chess is just to know how to play it and to understand an alien culture is just to know how to act within it. Again, the identity of the constituents of a culture is given in precisely the same way that the identity of chess-pieces is conferred by the rules. A bishop is identified solely in relation to the rules of chess as is the move of a pawn or knight, and the same is true of any particular cultural institutions. It makes no more sense for the sociologist to claim superior understanding to members of a culture than it would for an analyst to say that knights ought to move diagonally. Similarly, cross-cultural comparisons are as nonsensical as the attempt to compare being off-side with being leg-before-wicket. So certain is Winch of this conclusion that he is even willing to accept a formulation of it from one of his critics: 'An agent's action "is identified fundamentally as what it is by the *description* under which he deems it to fall"' (emphasis added).[22] Moreover, it is not just in *describing* actions that actors necessarily have priority, but also in their evaluation and assessment. The rules of chess constitute the nature of knights and bishops and they also determine what counts as a successful move. The integrity of language-games applies as much to evaluative as to descriptive statements.

An oft-cited analogy employed by Winch to express his conception of the social sciences is that a student of society is not like an engineer examining the mechanical processes which make a piece of machinery function, but like a student of engineering acquiring an engineer's knowledge through working with a skilled colleague. Native members of a culture are expert in their social knowledge and all that the sociologist can aim for is the same degree of expertise. However, the native and the skilled engineer may possess their expertise tacity, as 'unreflective understanding'; the sociologist can aim to spell out this understanding but even when it is thus elaborated the understanding can only be good for its original purpose. And finally, of course, only members of cultures can authoritatively specify what purposes are to be recognised. The non-scientific nature of some traditional beliefs about the natural world can set purposes for knowledge which differ greatly from the rather instrumental purposes for which we assess knowledge, but no cross-cultural evaluations of these purposes can be made. Our science fits our purposes just as well but no better than alien beliefs, say about witchcraft, fit other cultural purposes.

Without following Winch in regarding every culture as hermetically distinct and literally incomparable with all others, it is possible to argue for the inevitable distinctness of sociology from natural science on the basis of the role of meanings. Such an argument is developed by MacIntyre in the course of his (previously cited) critique of Winch and the thread of his contentions can usefully be followed here. This position is, however, not specifically original to MacIntyre since it lies behind the whole series of claims about the dependence of social science on interpretative understanding or *Verstehen*.[23]

Fundamentally, the suggestion is that Winch is strikingly correct to insist that people's own descriptions must be used to characterise their actions because their conception of what they are doing actually identifies the action. That is, people inevitably provide the best descriptions of their actions not just because they are, so to speak, on the spot, but because it is their understanding which confers an identity on those actions. This too is the claim advanced by social phenomenologists. For example, they contend that what it is to be deviant – to be a drugpusher, say – can only be appreciated through a grasp of the deviant's view of the situation. An external view of deviants' conduct is likely to make it appear senseless whereas a grasp of the meanings deviants attach to their own actions reveals their sensefulness. Thus, it is suggested, unlike natural science where external measures and observations are adequate, sociology depends for its initial level of descriptions on an an interpretative grasp of people's own meanings. The practical upshot of this difference can be pointed out at the level of social policy: unless legislators understand the motivational orientation of deviants it is impossible to introduce wise laws and policies. For instance, what lawmakers consider to be a punishment may not actually be regarded as a privation in the deviant subculture and any laws instituted on this basis will inevitably be ineffective.

This view of the importance of members' meanings is not as radical as Winch's position. For Winch, cultures both supply the correct descriptions of actions and the correct explanation and evaluation of their occurrence, while MacIntyre wishes to apply cross-cultural explanatory principles once correct descriptions have been obtained. When one has, for instance, established that some social institution in a foreign culture is concerned with deciding questions of landownership or responsibility for the care of children then it is permissible to compare these institutions with the apparatus employed in other cultures for the same purpose. Moreover it is possible to compare the efficiency of these various institutions and even to generalise about correlations between, say, 'legal' institutions and the hierarchical distribution of ownership rights. Against Winch he insists that two undeniable facts contradict Winch's notion that cultures form hermetic units. Firstly the practical possibility of translation means that some degree of compatibility must exist between the descriptions offered in various languages. Secondly, changes which occur within a culture's beliefs (such as the overthrow of witchcraft beliefs in the West) show that, in principle, members are capable of criticising their own culture. Accordingly the analyst cannot be logically obliged to accept the culture's claims either.

On this view, therefore, sociology differs from natural science in that the social world is pre-interpreted by its members. These interpretations serve to constitute the social world so that the analyst is not free to describe the phenomena in whatever terms the discipline permits. A sensitive analysis cannot gain knowledge of the social world except through the categories identified within it by the native members. However, once this initial level of descriptive adequacy has been obtained, analysts are free to apply any explanatory devices to whichever questions they see fit to ask. Thus, although there is a major difference between the natural and social sciences, its analytical significance is limited: in both cases it appears that what you do with observations and descriptions once acquired is up to you.

A different sense in which it might not be entirely up to you is related to moral and normative questions. Rather than try, as Habermas does, to erect an emancipatory image of sociological knowledge on epistemological grounds, one can introduce a broadly moral argument for the special role of such knowledge. Considerations of this sort have often been ascribed to Marx, but they are more unequivocally associated with the 'Critical Theorists' like Marcuse and Horkheimer whose work attracted much attention in the sixties.[24] Their fundamental contention is that in the course of history both societal organisation and human knowledge (and cognitive dispositions more generally) evolve in a one-directional fashion. Whilst the specifics of the connection between societal change and the development of human capacities have not been agreed in detail nor the question of the causal priority of one over the other, the general upshot of this connection is accepted. The chief point is that any supposed universal law of human conduct (such as, for a facile example, that people are universally greedy or that the rich get richer ...) may actually be a statement which is true of only the present stage of history. Accordingly, and this is the central claim, any such law may actually be a reactionary force if it commands wide assent. And, clearly, this is particularly worrisome if human beliefs actually play a large part in determining the course of history.

On this view social scientific knowledge has a very important moral and political dimension which is quite distinct from the position of natural science. The 'law' that capitalist production operates by the extraction of surplus value is not a brute, unalterable finding of scientific investigation. Instead it is the statement of a temporary state of affairs which, if widely known about, could lead to changes in political opinion and the eventual overthrow of capitalist production and thereby the negation of that 'law'. This point has largely been made by analysts on the political left but it can stand as a general argument against naturalism. Generally, at least some findings of social science could be transfigured by people coming to know about those findings so that sociology has a political obligation built into it. Unlike Habermas, proponents of this view base their claims not on an abstract epistemological scheme but simply on the moral and political character of human beings. If we felt the same about the objects of natural science (as perhaps we are beginning to in the biological realm) similar considerations would apply to that knowledge.

Additionally there are two, less politically-radical arguments for the distinctness of social knowledge which derive from broadly moral considerations. The first concerns value neutrality in social scientific terms.

It can readily be argued that many terms currently used in describing societal conditions are value-laden, for example the appellations democratic, sexist and fascist are used judgmentally as well as descriptively. The H-D response to this difficulty is to seek to separate the two components of these terms. It is acknowledged that all analysts will have their own personal values but these are taken to be contingent, unsystematic commitments. Because of their irregular variability from one analyst to the next, values should be removed from all descriptive terms. In a celebrated article, C. Taylor has argued against the possibility of such a separation on two grounds.[25] For one thing it makes a mockery of people's values: if people's values were as arbitrary and independent of political and social facts as the H-D view suggests then all debate would be thoroughly anarchic and all social science redundant since it could affect no-one's opinions. People who can adduce no reasons to support their values are not viewed as unusually committed, they are regarded as obsessive. Accordingly, the H-D view of sociology seems to imply that all social actors are likely to be obsessively attached to, and immovable from, their values.

His second argument is that the H-D position is incorrect in allowing only that values may affect facts but not vice versa. Taylor contends that some factual claim such as that totalitarian governments can only be sustained in advanced capitalist societies by the extensive use of violence cannot but have an evaluative upshot. Other facts about the social advantages of such governments might also be brought forward, but in the absence of these the facts inevitably lead one to a negative evaluation of totalitarianism.

Moreover the interaction between facts and values can lead to sustained differences concerning social issues. A democrat from East Germany would defend the democracy of that country by appeals to facts which make Western states appear relatively undemocratic (such as the availability of employment and low salary differentials) and a West European would argue in a contrary direction but in a similar fashion. For Taylor social science simply cannot be segragated from moral and political concerns. Therefore there is no way in which value neutrality, the apparent price for assimilation of social knowledge into a naturalistic science, could be achieved even if it were thought desirable.

The final argument in this section derives from questions about the freedom of the will. Developed both by analytic philosophers and by existentialist authors, it commences from the possibility of individual choice. The suggestion is that the existence of choice demonstrates that sociological generalisations can only be aggregations of past choices and cannot have predictive or causal force since at every juncture there intervenes a human decision or choice. At most, prior events serve to generate a situation in the face of which a choice has to be made, but there is essentially something distinctive or ineffable about the act of choosing. A variety of defences for this distinctiveness have been put forward including the status of mind as a separate category of existence from physical matter and the fact that any generalisations, being the product of human minds, cannot be superordinate to other minds. More mundanely, it has been argued by Strawson that our common practice of attributing blame and responsibility (which features both in everyday life and in the courts) simply requires that we assume that people are uniquely in charge of their own actions. If we regarded all action as caused these institutions would crumble. On this view it is the distinction between moral agency and non-moral causation that divides the natural sciences from social knowledge and which disqualifies a naturalistic social science.[26]

Methodological objections to a naturalistic sociology

A number of related points arise under this heading which can be traced to a body of writings commonly termed ethnomethodology. In a sense these are also pragmatic arguments against naturalism since they deal with methodological obstacles to its implementation. But the obstacles are claimed to be of a singularly insurmountable sort and it is suggested that study of these obstacles illuminates an alternative strategy for sociology. Ethnomethodology can be summarised as the logical extension of MacIntyre's suggestion that people are fundamentally their own sociologists. Ethnomethodology maintains that people are not just generally endowed with social knowledge which describes their surroundings; rather people constantly exercise their social knowledge and are forever theorising about each other's actions.

The first thrust of this viewpoint is against empiricist sociological methods. In obtaining interview replies, in extracting factors from open-ended interview responses, or in the interpretation of any data, it is suggested that the analyst calls on a great deal of unexamined common-sense knowledge about people's typical actions, intentions and dispositions. So many uninvestigated and unexaminable skills are employed just in, say, conducting an interview and these skills contribute to the outcome in such oblique and unknown ways that any purported sociological data are a confused amalgam of information about the respondent and the product of the interaction which yielded the data. There is simply no way of discerning how much of the data can be ascribed to these different effects, so any conventional sociological work actually trades on data which are flawed to an unknown and unknowable degree.

However this negative point about the uncertainty of any single piece of social information also has a positive component. In everyday life no interaction is entirely unambiguous and no meaning can be specified exhaustively - it is always possible that our most trusted friends are deceiving us for reasons of which we are ignorant - but this openness allows for nuance in speech and for creative uses of language. Thus all utterances are actually interpreted in the light of the context in which they occur, and it is only our skilful assessment of what those words are likely to mean in this context that allows us to get by in the great variety of novel contexts in which we constantly find ourselves.

Some rather bizarre investigations have been carried out to show just how much a judicious assessment of the context contributes to the understanding of utterances.[27] For example, subjects were asked to 'help out' with a training programme for supposed student counsellors; they were asked to seek some advice of the counsellor but only through questions which admitted of a simple yes/no answer. Actually 'student counsellors' were experimenters who gave a standard, randomised series of answers to the questions. When the subjects were asked afterwards what they thought of the counsellor's replies they were all found to be able to construe some coherent, overall sense out of the series of yes/no answers they had received. By invocation of obscure motives which might be attributed to the counsellors and by reference to possible peculiarities of the context (like, perhaps the counsellor had received too little initial information on the topic under discussion) the subjects managed to make sense of their interaction. For ethnomethodologists this mutual dependence of meanings on the context and on the interactants' motives is a general property of social life - they term it indexicality. It is commonly suggested that natural science is importantly non-indexical in that the meaning of terms is subject to rigorous definition. In social life it is of the essence that meanings are not do defined.

This realisation has a gloomy upshot for naturalistic social science since the supposed generality of any sociological statement will be undermined by indexicality. It is never possible to say that this Protestant family (as Durkheim suggested) has lower social solidarity than this Catholic family because of religious differences. The difference might always be attributable to some other particulars of these families or to some features of the circumstances in which the observations were made. There is thus no solid basis for any generalisation since they are all founded on some unspecified judgement that, despite indexicality, the differences between these families and those others can be attributed to varying religious affiliations.

For ethnomethodologists, it must be the case that any purported sociological generalisations are based on unexaminable and essentially arbitrary analysts' assumptions. The alternative course which they propose is to study the anatomy of the procedures which people use in overcoming indexicality and reasoning in everyday situations - that is, to study people's (ethno)methodology. This may appear uninteresting to members themselves - much as linguistics may appear irrelevant to competent speakers - but it is the only avenue for sociology. One procedure that people use is termed the documentary method. That is, people regularly employ certain conditions as documents of other states: actions and appearances may be taken as indicative of mental conditions and posture may be taken as an index of certain kinds of intention. People use the documentary method to engage in sociological theorising about the meaning of each other's actions. Because of indexicality ethnomethodology maintains that analysts cannot do better than ordinary people (or members' as ethnomethodologists tend to term them) themselves. Indeed, in an important sense there is no better to do, since people's assessments of others' actions are carried out for that moment alone. To seek to give a definitive sociological portrayal of any situation would be a useless interruption and ossification of social life.

Another characteristic of members' methods is that everyday reasoning is essentially *ad hoc*. There are two facets to this: firstly since the words used in everyday speech are not subject to rigorous definition there can be no set creteria for establishing what other people mean or are talking about. If you are trying to describe someone whom you met at a party there is no single 'correct' description of them. Instead your description is taken as adequate when other people claim to recognise the description. In the same way, all descriptions are taken to be *ad hoc*; their adequacy can only be adjudged in practical terms.

The second sense of '*ad hoc*ness' is more complex. As an example, in some way we all know or assume that people's actions are related to their motives. Yet in the last instance the only evidence we have about those motives derives from people's actions (including speech acts) and it is alway possible that those actions may be a front for even deeper motives. In many cases our estimations of other people's motives are subject to *ad hoc* modifications in the light of their subsequent acts. And it is forever possible that new facts will come to light which lead us to revise these estimations. Therefore there can be no final check on our comprehension of the actions and motives of other people. Instead we must make some practical decision about what others are likely to be up to. All social life depends on such decisions, but they are all equally *ad hoc*. Thus, there appears to be an everyday working assumption that people's acts and motives *do* possess a true identity, and this assumption is complemented by a range of devices for *ad hoc*ing when provisional identifications of motives and actions appear to be erroneous. However, ethnomethodologists take the view that social actions have no intrinsic identity – their only identity is the one foisted upon them by decisions made in subsequent interactions.

Consequently, ethnomethodologists deny that any firm causal connections can be made between social facts since all facts are liable to re-interpretation. Of course, members may come to assert causal connections – priests may assert that evil causes delinquent behaviour – but these assertions are found only on an *ad hoc* basis and may well come to be revised. Thus, while people seem to claim something like naturalistic knowledge of social life, ethnomethodology argues that this is an appearance sustained by *ad hoc* reasoning. Since causes can be re-interpreted and re-construed there is no prospect for an enduring social science.

Finally, since these conclusions may seem to have something in common with Winch's ideas it should be pointed out that ethnomethodologists do not accept that social life is actually regulated by rules. It is claimed that rules too are interpreted in an *ad hoc* way such that it is not possible to say that obedience to a rule preceded conduct which appears to conform to that rule. People can invoke rules to justify actions which were not originally carried out with respect to those rules. Someone caught with unpaid-for goods in their possession may have been shoplifting or may honestly have forgotten to pay. Only a decision (founded on *ad hoc* reasoning) determines whether they are considered to have been complying with society's rules or not.

Conclusion

The H-D empiricist theory provides an impressive and compact account of the supposed essence of scientific knowledge, and for this reason can readily be supposed to have exercised a large influence on how social scientists have designed their research procedures. However, both practical and philosophical objections have been raised to these procedures. The assimilation of sociology to naturalistic science has been resisted on a variety of grounds – including the special properties of the subject matter of sociology (i.e. the moral and political facets of human life or the pre-interpreted nature of social reality), and the distinctive emancipatory orientation of social knowledge. Before the significance of these various objections can be assessed it is vital that the accuracy of the H-D account of scientific knowledge should be examined lest these objections to naturalism should be directed towards a distorted image of it. In the next chapter we shall see how revisions in the empiricist account of natural science affect the force of the objections.

CHAPTER 2

The naturalism of the new philosophy of science

Facts and theories in the new philosophy of science

In this chapter I shall indicate the principal revisions which the H-D view of science has undergone and investigate the extent to which they serve to meet the objections to a naturalistic social science reviewed in Chapter 1. In particular, it will be important to assess how much the anti-naturalist arguments were directed at a falsely positivistic interpretation of science. To some extent the discussion will cover material which has been argued by other authors and I shall accordingly present a fairly brief account.[1] The emphasis will be on substantive and procedural issues not on questions of philosophy as such.

The central claim of the H-D view of scientific knowledge was that facts are primary and this primacy was emphasised in the account of Semmelweis' medical researches. Facts have logical priority and they can be incontestably established which, in turn, is the guarantee of sure and cumulative knowledge. For some analysts this led to an almost exclusive attention to the role of facts. Theories and hypotheses were seen as, at best, collections of facts and, at worst, tempting distractions from the business of gaining knowledge. Such extreme insistence was uncommon, however, and many writers acknowledged that theories provided valuable re-statements of facts and that hypotheses speeded up the growth of knowledge. But attempts at formalising this more liberal view ran into difficulties both because of the significant role of distinctly theoretical knowledge and because of a lack of straightforwardness in the realm of facts. The archetypal fact is something immediate and publicly observable: the falling of fruit or the colour of chemicals placed in a flame. As science progresses there are facts which deviate from this desirable pattern, facts which occur only under bizarre conditions or with peculiar instruments.

In general there are three ways in which facts depart from the ideal mould, which I shall call remoteness, shiftingness and universality. The obvious rule of thumb for identifying something as factual is that it should be observable. Remoteness often stands in the way of observation since an effect may for example only be generated under an enormous pressure in total darkness, so that no-one could get near it to see it, or the phenomenon may only be 'visible' through a microscope, where its status as an observable is thoroughly dependent on our assumptions about how microscopes work. This problem is naturally intensified in the case of an electron microscope where the image 'seen' is technically produced. Thus facts cannot be a certain touchstone because they are not unquestionably available to public observation. Despite our normal confidence in microscopes we can no longer argue in principle that there is a category of distinctively factual, observationally-based knowledge.

Shiftingness is closely related to remoteness. Certain phenomena, such as electrons, at one time had a conjectural existence and were not straightforwardly factual. However with their increased acceptance as fundamental to physics and chemistry they have come to take on a factual status, as is witnessed by the confidence we feel in electron microscopy. Since the objects of scientific knowledge can pass from a conjectural to a factual status it is clearly impossible to maintain that there exists a stable factual realm which is observable and acts as a guarantor of certainty.

The final attribute of factual knowledge is in many ways more crucial than the other two since it not only makes facts appear potentially uncertain but gives them a partially theoretical character. The archetypal fact might be thought to be a descriptive statement of some singular object, such as that 'this chemical is green'. But this example is actually misleading since many factual claims are of the same kind as 'all electrons have a negative charge'. Although intended as an ordinary factual claim, this latter statement purports to refer to *all* electrons and cannot therefore, in principle, have been based on exhaustive observation. There is undeniably something hypothetical about the claim that this statement is true of all electrons. There is room in H-D philosophies for claims of this sort if they are definitional: an example might be 'all birds have wings' (although even here the existence of flightless birds throws some doubt on our confidence in the immutability of definitions).[2] But if they are empirical claims a problem arises since it must be logically possible for the claim to be frustrated: some as yet undetected electron might not have the usual charge. This problem is known as the problem of induction. It has often been posed as a philosophical conundrum (how can we justify our belief that the sun will rise tomorrow?), but people have rarely taken the implied anxiety to heart. The true significance of the problem is not what it reveals about the uncertainty of tomorrow but concerns the inadequacy of the traditional empiricist conception of facts. It demonstrates part of the hypothetical nature of factual knowledge.

Perhaps sensing these three imperfections of facts in relation to the demands placed upon them by H-D empiricism, philosophers have sought to maintain the separateness of facts by describing science in terms of two languages. But this distinction between observation and theory languages (encapsulated in the 'instrumentalist' view of theories described in Chapter 1) does not meet the requirements. The principal innovation of the new philosophy of science occurs precisely at this point, for it maintains that there is no defensible theory/observation distinction. All observations are interpretations, dependent on the state of our beliefs. Thus 'factual' observations are intimately associated with our theoretical commitments.

In some ways this assertion appears less satisfying intuitively than a view which accepts a theory/observation distinction. It might be argued that observations are given by the light pattern transmitted to our visual apparatus and that we then interpret them in a discrete act. This will not do, however, since we see what we have learnt to see: we cannot but see the world in terms of trees and cars. We never see any uninterpreted patterns of light. Similarly, we do not see coloured objects in the sky, we see stars and the sun and moon – and astronomers see particular stars and planets. Our beliefs about what there is to be seen shape the possibilities of what we can see. Naturally, there are relatively neutral observations but there are no preconception-free observations on which to ground an empiricist science.

A complementary tenet of the new philosophy of science is that theoretical knowledge is under-determined by (relatively) factual knowledge. Since pre-existent beliefs are required to interpret any visual stimulus it follows that alternative theories could be based on the same stimuli. No level of informational input can uniquely determine what theoretical interpretations it supports. Thus theories are under-determined: they are constrained but not fully specified. Such inter-dependence of facts and theories is taken to be an entirely general property of knowledge and thus upsets the hierarchy established by H-D views of science. Indeed, it virtually inverts the previous ordering since the acceptance of facts is now treated as being relative to the adoption of theoretical commitments. And this result penetrates directly to the heart of the dominant theory of science for it is no longer possible to accept that science grows in a linear fashion – there is more to knowledge than the incremental addition of new facts – nor is there a ready means for assessing one theory against another because the criterion of factual correctness is relative to the theory one holds.

The difficulty raised by the possibility that the facts observed are dependent on one's theoretical persuasion is termed the problem of relativism. If facts are relative to one's theories then there is no way of determining whose knowledge is best or most accurate; worse still it becomes meaningless even to ask which theory is correct. A very small minority of philosophers, such as Feyerabend, have exulted in the prospect of relativism, suggesting that it opened the way for the simultaneous development of many alternative viewpoints and that it encouraged a form of cognitive liberty, but for the mainstream philosopher of science it presents an annoying problem.[3] Annoying problems (like the problem of induction) need not create deep anxiety, however: the possibility that scientific knowledge does not grow or that theories cannot be compar atively assessed is not generally seriously entertained by philsophers. Relativism provides a riddle to be solved not a disproof of science. It prompts new attention to the theory of science.

Comparing and modifying belief

The issues raised by the new, more complicated view of the relation of facts to theories have so far been treated only at an abstract level. Their practical importance can best be assessed by inquiring how scientists are to go about evaluating various factual and theoretical claims and beliefs. This can conveniently be done if we consider a scientist who ascribes to some theory: for example, that rock strata are consolidated by the influence of heat. This scientist can be said to have internal and external problems of evaluating and modifying belief. By internal problems I mean the difficulties encountered in the attempt to develop the theory. The scientist may find that some strata (like chalk) are of a substance that would decompose under the influence of heat and thus could not have been consolidated in the way the theory anticipates. Philosophers generally speak of such internal problems as anomalies. If facts were primary any genuine anomaly would discredit the theory. A similar result would be produced by Popper's suggested method of falsification. In his well-known work on the logic of science he proposed that theories, being conjectural, should be continually tested against the observable world and retained only so long as they were not falsified. For Popper, a good theory was one that was open to falsification but which continued to pass all tests to which it was subjected.[4]

The strategies suggested by these philosophers for coping with such problems neither reflect what scientists do with internal problems nor what they should do. It has already been remarked that philosophers, when confronted with the anomaly of relativism, do not give up their belief in science but try to comprehend the anomaly. A similar approach is adopted in science. In the first place, it has to be shown at what precise point the anomaly came about. Faced with an apparently anomalous claim about a stratum of chalk, the scientist may doubt whether the observation was correctly made (the rock may look like chalk but is it really . . . ?) and whether it is relevant (does the chalk make up a stratum or merely a vein which formed in a different way?). Only once these doubts have been met, and in the case of complicated detection apparatus for submicroscopic entities these doubts can be multiplied enormously, will the scientist have to acknowledge an anomaly. And the response then will be to treat the anomaly as a puzzle. The scientist will probably modify the theory to make it more specific by saying, for example, that less heat may be needed to consolidate chalk than to decompose it, or that under pressure chalk might not decompose even at high temperatures. One prime task of the philosophy of science must therefore be to describe the strategy appropriate for dealing with internal problems of belief.

Turning now to external problems, these relate to the question of when a scientist should switch allegiance from one theory to another or which of two new theories is to be accepted. Popper's falsificationist answer is that scientists should submit each theory to the same tests and accept the one that is not disproved. But we have already seen that these disproofs may appear to adherents of the theory as mere anomalies and, worse still, if the scientist employs a number of tests neither theory is likely to come out totally unfalsified. Thus, our geologist might also know of a theory that strata are formed by chemical precipitation. This theory fits well the case of chalk but not insoluble quartzy kinds of rock. Both theories fit some cases very well and can be modified to cope with apparent anomalies. The philosophy of science must aid the scientist in discriminating between the alternatives.

The most popularly known account of how science operates in the light of these philosophical revisions has been provided by Kuhn in his book, *The Structure of Scientific Revolutions*,[5] He suggests that the primary units of the development of science are schools of thought associated with one overall theoretical perspective. These units he calls paradigms. Paradigms supply a central substantive theory (such as that continents are located on mobile but enduring plates of the earth's crust); exemplary methodological procedures (for tracing the movements of plates through geological and geophysical techniques); and the basis for a loose social group (people working on the theory of the earth's crust tend to read the same journals, attend similar meetings and work in similar departments). According to Kuhn, the paradigm as a whole commands wide allegiance and anomalies are tolerated on the assumption that their solutions will be found in the long run. Most scientists work on small additions and improvements to the paradigm. This may require great skill and perseverance but leaves the fundamental outlook of the paradigm unaffected; such work is dubbed normal science. From time to time dissatisfaction with the paradigm mounts and the accumulated anomalies are used as grounds for overthrowing one paradigm and replacing it with another. These are periods of scientific revolution. The replacement of the Newtonian theory of physics by Einstein's relativity and the associated quantum theory of matter is probably the best-known example of one paradigm succeeding another.

Many historical writers have found much to identify with in Kuhn's account and it has attracted a great deal of attention, but for present purposes it is important only to ask what the Kuhnian analysis tells about scientists' investigative strategies. Firstly, Kuhn does not claim that accumulating anomalies overturn a paradigm; rather anomalies are exploited by the forces advancing the revolutionary coup. Perhaps solutions would have shortly come about within the old paradigm or perhaps the problems would have remained intractable. Anomalies are crucially the *occasion* for a revolution not its *cause*. Secondly a new paradigm typically makes a virtue of supplying answers to these anomalies. That it should be superior in this weak sense is a condition of the plausibility of the revolution but Kuhn does not claim that the new paradigm is better in any wider sense. Characteristically, there is no net growth in knowledge since some of the low-level knowledge associated with the former paradigm is lost in the transition. With the demise of alchemy, for example, much knowledge about mixtures of chemicals passed into oblivion. The extent of the change may be so great that there are very few criteria by which the paradigms could be compared. Thus although Kuhn allows that the world's real constituents constrain our beliefs by generating anomalies within paradigms he does not accept that the world's constituents can discriminate between paradigms. So many historical and personal forces impinge on paradigms that it makes no practical sense to talk of one paradigm being better or truer than another and even less sense to think that paradigm successions mark steps towards true knowledge. Finally, Kuhn insists on the role of dogma in science: since to work in a paradigm one must accept its central assumptions and procedures the majority of creative scientific activity is carried on within tight cognitive constraints. Unlike the Popperian scientists who tirelessly put every belief to the test, the Kuhnian scientists can do creative work only when they leave certain beliefs inviolable. For Kuhn paradigms operate very much in the way that cultures do for Winch:

each supplies its own criteria for meaningful activity and there are no over-arching or transcendental standards of comparison. Kuhn refers to this absence of comparability between paradigms as 'incommensurability'.

Kuhn's emphasis on dogma and anomalies was embraced by many historians of science (and of other disciplines like art) but it did not serve philosophers' requirements – if anything it made the problem of relativism worse by making it more plausible. While some philosophers denounced his work, Kuhn's arguments were taken seriously by Lakatos who approached the debate from a Popperian standpoint.[6] Lakatos accepted that a theory of science had to describe scientists' methodological strategies over time and he accepted that these were based around paradigm-like commitments which he termed research programmes. He elaborated a theory of research programmes which attended both to scientists' external problems (corresponding to choice between paradigms) and internal problems (which Kuhn had left rather vague).

To develop this theory it was necessary to identify the structure of beliefs and commitments which comprise a research programme. Essentially the structure consists of a hard core of deepest theoretical commitments which could not be altered without the research programme ceasing to exist. This core is surrounded by a protective belt of beliefs which scientists are more or less willing to change. The only interface that the research programme has with low-level, factual-type statements is on the perimeter of this protective belt. Data are provided from experiments and observations through interpretative theories which develop alongside the explanatory component of the research programme. These interpretative theories cope with the 'hypothetical' aspects of facts described earlier: specifying how electron microscopy functions and so on. Lakatos then aims to derive rules which would lead to research programmes developing as the best available models of the natural world and asks whether scientists actually follow these rules.

The fundamental rules are known as the positive and negative heuristics: these encourage extension of the programme in a progressive direction and discourage amendments of the hard core. A research programme is said to be progressive when the theory is being extended in ways that are confirmed by observation and experiment and regressive when facts are being explained after the event by adjustments in the protective belt. Returning to our primitive geologist, a Lakatosian would describe the research programme as progressive if, say, the scientist expected that during consolidation strata would lose moisture and thus crack leaving fissures and such fissures were then detected as expected. It would be regressive if an anomaly were detected – like collections of flint – which none of the protective belt of theories predicted and for which a special theory had to be introduced.[7] A programme is regressive when alibis have to be found to explain anomalous occurrences.

From these kinds of consideration rules of method can be deduced. Thus, if there are two theories, choose the more progressive one if they are equally powerful in other respects. This rule should take the arbitrariness out of choices between paradigms. Other rules include: adjust the protective belt to protect the hard core from anomalies, which rationalises Kuhn's normal scientific dogmatism; and if there is no competing research programme stay within your own even if it is not progressive, a rule which rationalises people's reluctance to give up falsified theories. Lakatos' theory of science maintains Popper's emphasis on testing belief whilst allowing for the complex structure of belief systems. It acknowledges that scientists behave in many of the ways that Kuhn indicated but insists that they behave in this way because in the long run it is rational to do so.

Although philosophers acknowledge that there are further elements of this theory of science which still need to be specified, for example the precise distinction between the hard core and the inner protective belt, it is the view which attracts the most support in the new philosophy of science. None the less there are three areas in which it is open to criticism and where no internal revisions appear likely to avert these criticisms. I shall, however, only mention these here saving the main discussion until Chapter 4. The points are: the precise details of the mechanism for deciding between alternative theories; the relationship between Lakatos' theory and the history of science; and difficulties in coming to a practical but unambiguous interpretation of the rules of method. Setting these questions aside for the moment the most marked characteristic of this form of theory of science is its transfer of emphasis, from particular beliefs, to specific methods for modifying belief as the central characteristic of science. That is, Lakatos does not make a belief in goblins or in unobservable mental states unscientific. Rather, a belief is unscientific if it is not treated as part of a continually assessed research programme. If facts are regarded as supreme then belief in the right things is symptomatic of science, but with research programmes as central it is belief for the right reasons. An unrelenting rational appraisal of belief is the defining feature of science. Thus, for example, Newton's (infamous) belief in alchemy is no longer a bizarre departure from rationality since he may well have subscribed to it as the best theory of the constitution of inorganic substances available at the time.[8]

With this shift to method and away from substantive beliefs it becomes natural for science to aim at successively better theories or research programmes and associated models rather than at a final truth. Certainly it is pragmatically inadvisable to consider that one's beliefs may be flawless. This lack of concern with truth raises traditional philosophical questions about the relation between our knowledge and the contents of the real world. Theories like that of Lakatos which are agnostic about the real world are often termed conventionalist since every research programme is held to be a conventional depiction or model of some portion of the universe. Other philosophers consider that conventionalism has its dangers since it can lead to an instrumentalist conception of all knowledge. These latter philosophers - generally termed realists - argue that unless we suppose that successive theories disclose more and more real knowledge we are implicitly admitting that the scientific method (or Lakatos' rules) is merely a convention or language-game which has no inherent superiority to other conceivable knowledge-games.[9] Realists typically argue that without a commitment to the reality of the causal forces described in our theories there is no difference between a prediction generated by our theory and one generated by a rule of thumb. They acknowledge that some 'laws' of science are purely formal and have no reference to explanatory forces (such as the laws of motion which permit calculation of trajectories without saying why matter behaves in the described way). These laws are described as 'reticular' theories and are regarded as the exceptions which prove the rule about what is expected of theories. Generally speaking, good theories imply the existence of a real causal mechanism.

These two contemporary philosophies undeniably differ, but in terms of their consequences for the conduct of (social) science they differ much less than one might anticipate. Briefly put, and I shall return to this issue, the imputation of a form of realism - or perhaps as-real-as-our-knowledge-currently-allows-ism - appears to be a common concomitant of sustained work within a research programme. Similarly, scientists' commitment to the real nature of some causal connection cannot be so strong that it prevents them from changing their mind. Realism and conventionalism are not in this sense exclusive alternatives.

The new philosophy of science and social science

The new philosophy of science can be said to have both direct and indirect implications for the assessment of the scientific standing of sociology. Considering first the direct implications, the rehabilitation of unobservable entities and a more liberal conception of facts permits sociological notions like unconscious wishes, class consciousness and norms to become the object of scientific attention. It does not licence any particular theory as scientifically superior but does demonstrate that theory involving these conceptions is not unscientific. Moreover, the continued existence of competing Marxist and liberal theoretical perspectives can be interpreted not as a mere ideological clash but as the conflict of alternative research programmes. And the characteristic strategies found in these perspectives (for example, modifications of Marxism to explain the passivity of the Western proletariat) can now be viewed not as evading the facts but as making adjustments in the protective belt to shield the hard core. Providing that Marxist theory is being successful in some area, perhaps in the explanation of neo-colonialism, it is perfectly scientific in Lakatos' terms to be making *ad hoc* modifications to other aspects of the research programme.

Equally, the lack of mathematisation in sociology is no longer a fatal flaw in its scientific credentials since reticular theories are commonly considered less desirable than causal ones. While little mathematical expression can be given to the theory of the class struggle, a theory of class antagonism can readily be interpreted as predicated on the causal underpinnings of capitalism as an exploitative organisation of production. In this way many of the general features of sociology which are often held to be unscientific can be shown to be in conformity with Lakatos' stipulations. Other strands in the anti-naturalist case can also be picked up. The new philosophy of science acknowledges that experience is necessary for carrying out scientifc work and for 'seeing' the facts of science properly. Hermeneutic forms of interpretation are no longer unique to the social realm either, since the under-determination of theory by facts requires a creative and interpretative style of thinking in natural science too. Science is still held to be a matter of method but now the method involves imaginative response not automatism. In this way Habermas' type of distinction between the interests of natural and social science begins to evaporate. According to Lakatos all theories and all facts are instrumental in the sense of being conventions so there is nothing remarkable in science's instrumental interest.

Social science's concern with meaning is also affected since paradigms or research programmes depend on developing elaborate systems of thought with only peripheral anchors in the world of facts. It is accepted that these edifices of invented meanings cannot be reduced to facts but this need not rob them of an objective comprehensibility. Such meanings must be open to scientific understanding and if they can be understood in this way then why not other meanings too? Moreover, the use of models and the invocation of ideal types begins to resemble the method of a conventionalist science.

The final direct impact of these revisions in the philosophy of social science is on the arguments of ethnomethodology. It was claimed that an ethno-method had been recorded which differs from the scientific method. This ethno-method was characterised by *ad hoc*ness, indexicality and the documentary method; attributes claimed to be absent from scientific method. Yet *ad hoc* modifications in the light of a mis-match between one's theory/assumptions and the context is just what Lakatos recommends for the response to anomalies. Social objects are supposed to have indexical (i.e. contextual) descriptions whilst scientific objects can be given literal descriptions, yet Kuhn and Lakatos agree that objects are indexical across paradigms.[10] Finally, actors in the social world are supposed to impute reality on the basis of mere documents: actors make out that, say, inner feelings lead to facial expressions whereas the ethno-method proceeds from the expression to the supposition of the feeling. But scientists impute the existence of sub-atomic particles on the basis of a trace on a pen-recorder or a bubble chamber. The use of documents to record a property or thing which then accounts for the document is not peculiar to social life; it is equally the method of conventionalist science.

This discussion is not offered as an invalidation of ethnomethodological research, of which I shall offer a positive evaluation in Chapter 5. It is however a prime instance of anti-naturalist arguments drawing their strength from a false image of science and of the (perhaps undeserved) discredit which anti-naturalist researchers can lead themselves into. It also serves to demonstrate that the direct implications of the new philosophy of science are mostly for the self-image of social science. We shall now turn to the more substantively significant indirect implications.

Philosophy of science and the image of the social actor

A theory of social action cannot be deduced from the new philosophy of science but the theory of the actor as a rational decision-maker has taken precedence, both because it can be treated as scientific by Lakatosian standards and because the scientist in Lakatos' schema acts by taking decisions rationally.

The kernel of this theory is very straightforward: actors are presumed to have sets of beliefs and of wants, and actions are selected to satisfy their wants in a manner that their beliefs indicate is correct. It is argued that this calculus of wants and beliefs underlies explanations of action in terms of reasons and that this underpinning shows that reason-type explanations are scientific in the requisite sense. Hence if someone is asked to explain their action an answer is typically provided of one of two sorts – supposing that the action was attendance at a demonstration, the reasons might be either:

1 I thought the demonstration would influence the University's investment policy.
2 I wanted to change the University's investment policy.

The first case refers to *beliefs* about the University's response, the second to the speaker's political *wishes.* Yet if, in the first case, the speaker had not *wanted* to change the University's policy then the reason would not make sense. Neither would the second reason be good unless the speaker *believed* the demonstration might function to influence the University. Thus it is argued that all reason-type explanations refer to wants and to beliefs.[11]

Although it is a rather elementary theory as yet it does seem to accord with Lakatosian requirements. It can be tested, for example by seeing how people with similar wants but different information act. And the hard core is protected by supplementary notions, such as that people can have conflicting wishes or ill-formed beliefs. Moreover the theory can be developed progressively to examine complicated decision-making strategies. One example of progressive development is the 'laws' of supply and demand in economics where, on simple assumptions about the beliefs and wants of producers and consumers, the patterns of pricing and production can be predicted. Another instance is supplied by game theory which investigates the results of interaction between rational actors. The famous example of the prisoners' dilemma game shows how, under special circumstances, the straightforward pursuit of one's wishes leads to a paradoxical misfortune. Supposing there are two felons who are suspected of a major offence but who are apprehended together on a minor charge. Each is offered no punishment for the minor offence provided they give evidence against their partner on the larger crime. They are then locked in separate cells to think it over. Both will realise that it is in their own best interest to offer evidence. From the first prisoner's point of view (and calling the first X and the other Y): if Y stays silent, X can go free by giving evidence but will serve a short sentence by remaining quiet; if Y gives evidence, X will get a long sentence by remaining silent and a slightly shorter one for co-operating as well. The situation is symmetrical for Y. For each individual, whatever the other does it is better to offer evidence and so the result will be a slightly reduced long sentence for both – a less satisfactory (rational) outcome than could have come about had they somehow contrived both to stay quiet. Just because such rational games replicate real-world dilemmas it can be argued that rational models suit human action.[12]

At a specifically sociological level it might well be objected that the theory does not accommodate the many competing wishes that any individual may experience nor can it handle the uncertainty of social life. In principle, however, these issues can be treated through the concept of subjective probabilities. Subjects could be offered a series of graded choices between different combinations of outcomes so that both the value attached to particular outcomes and the person's perception of the likelihood of that outcome could be given a mathematical value. People would then be assumed to select the courses of behaviour which maximised this mathematical score. It is the claim of rational decision theory that it replicates actors' decisions, not that actors perform arithmetic calculations before deciding what to do. In a Lakatosian spirit, the contention is that this theory models behaviour best - particularly that actors do not act in opposition to the outcomes which they would most prefer. Admittedly, people may sometimes not act as the theory predicts but these episodes are often explained as due to changes of mind. And this represents a modification of the theory which can be tested, for we would not accept that an actor changed their mind unless there was, in principle, a detectable change in subjective probabilities. Other deviations from the model's predictions might be understood as irrationality induced by alcohol and so on. There is nothing unscientific to these amendments unless they are undertaken in an undisciplined manner. Its advocates claim that rational-decision theory is unrivalled in the explanation of human action and should thus be adhered to as currently the most progressive research programme.

Rationality and tautology in the explanation of action

An objection which is often raised against this theory of rational action is that it is tautologous. Briefly stated, the theory suggests that actors do what they prefer most but the only evidence about what they prefer derives from what they do. I shall later argue that there is a real methodological issue here but the problem is procedural not philosophical. In the meantime it will be worth quickly considering the allegation of tautology.

The theory would be plainly circular if we claimed that people only ever acted for their own gain or pleasure. We would then have to construe everything people did as pleasurable even if it involved a great sacrifice or discomfort. Instead of this we keep the range of wants and motivations very broad and open. Indeed one of the findings of rational decision theory could be the kind of wants that people have. In some aspects of the theory, particularly in its application to economics, one assumes that people do act largely for gain, but this is an empirical assumption. Some businesses are not run solely for gain and these deviate from the economist's model (economists sometimes refer to such activity as non-rational) but economics can only get by with this gross simplification because, in an advanced capitalist market, economic gain has to be a primary want. In a rigorously competitive situation enterprises which do not pursue this narrow end are in danger of ruin.

Since the economics of prices and supply readily lends itself to mathematical treatment a narrow, economistic concept of wants can be formalised far more easily than a conception of wants that attempts to include subjective probabilities over a large range of possible wants. This sometimes leads to the idea that rational decision making is intrinsically related to estimations of gain. No such implication is involved in the theory of the rational actor. In attempting to interpret actions, or series of actions, however, it is easiest to begin by seeing whether the action is compatible with likely wants and beliefs of the actor, and an obvious place to begin is with the postulate of desired gain. When the connection between the action and the actor's disposition is complex, as in some advanced games or in an elaborate market or when duplicity is involved, then the first presumed want is characteristically personal advantage. An analogy with natural selection in biological species may assist here. If we find a creature with some peculiar colouring or strange habits we first of all ask how this attribute may have been of evolutionary advantage. We do not rule out the possibility that the feature may be coincidental but as a first attempt we look for ways in which the feature is advantageous. If this procedure were pursued relentlessly then biology too might become tautologous, if the possibility that features came about other than as a result of direct evolutionary advantage were to be eliminated by definition. This procedure of looking first for an advantage is referred to as a regulatory principle of biological research. In an identical way it is a regulatory principle of economics that behaviours should be analysed for their potential for gain. (It may be added that a regulatory principle of certain forms of police work is to suspect criminality as a motivation in the first instance.) An over-zealous economism can make rational choice theory into a tautology but it need not do so.

Sociology may face a problem directly inverse to that of economics since individuals may have such a variety of wants that there is no viable regulatory principle. In the case of bizarre action we may strain our sense of possible wants to the limit. Indeed Laing's work on understanding schizophrenia as a language of covert protest takes behaviours which others have declared incomprehensible and shows a possible rational basis to them.[13] In this way the supposition of rationality need not be conservative as in the economistic case where everyone is presumed to seek material success; it can uncover unsuspected forms of motivation.

It should finally be added that rational decision theory does not depend on the dangerous procedure of having people declare their own beliefs and desires. These may be opaque to the individuals concerned. At least some part of the explanation of actions by beliefs and wants are open to public examination. Thus a parent beating their child may 'explain' the action by saying that the child ate its dinner too slowly. We are not obliged to accept this explanation since, without further amplification, we would not accept that this action really is generated by the cited reasons. We would suspect that some other reason was also present, such as concealed resentment of the child. In summary, therefore, the theory of action as the result of rational decision making is not tautologous because the circle of beliefs, wants and actions can be broken into and because we can each assess the validity of the connection between reasons and actions invoked by other people. In many cases the theory is not at all easy to apply without ambiguity but is is solid enough to suggest that in principle it is both scientific and relevant.

The model of the actor and social being

The model of social action espoused above is conspicuously individualistic: It is concerned with assessing the beliefs and wants of individuals and with explaining particular actions and patterns of behaviour as the outcome of imputed or attributed decision mechanisms. To this extent it does not appear significantly sociological. As Abrams has recently suggested, major components of the sociological perspective can be understood as a way of redressing the imbalance towards individualism in many political and ethical theories.[14] Sociology has, at least in part, been directed towards analysing the distinctively social attributes of human behaviour. To pick up on a few of the issues which concerned Durkheim: people in groups do not behave like simple aggregates of individuals; people's feelings as parents are not ones they have derived for themselves; and there appears to be a fundamental moral consensus (on which contracts and other arrangements between individuals can be established) which is not attributable to any specific persons. Even if the new philosophy of science does not prescribe a particular view of such phenomena, room must be left to cope with the explanation of this type of phenomena and specifically a technique must be available for making sense of the wants and beliefs with which individuals are primed.

The response of rational decision theory to distinctively social phenomena is to treat them as varieties of rules and contracts. Rational actors are capable of introducing their own and others' behaviour into the terms of rational assessment and this provides the foundation for understanding social regularities. Thus the two felons who were ensnared in the prisoners' dilemma could have anticipated their predicament in advance and arranged to stay silent. And if they feared that they would not be able to depend on each other they could have made arrangements for their agreement to be policed by other criminals. Through the introduction of this arrangement both actors would benefit. A number of ramifications can be shown (potentially) to follow from this state of affairs. Before organising subsequent jobs these convicts might insist that their new partners abide by the same rule; other law-breakers might follow their example; the original benefits of the arrangement could become confused with an attitude of defiance which enjoined suspects to remain quiet out of opposition to the police; and eventually the practice could become enshrined in a 'convict's code' which opposes 'grassing' and which ensures that anyone found grassing will be disciplined by members of the underworld. Once this procedure or norm is established the stakes in the prisoners' dilemma are altered since the imprisonment resulting from staying silent will be preferable to being freed and then informally revenged upon for turning informer.

Naturally, the whole paraphernalia of game theory is not required to spell out this line of thought on the convict's code which is familiar from any detective serial. What is significant is the way that norm or social rule can be derived from and understood in relation to actors' decisions. Whether an agreement is set up rationally or by chance and whether it is adopted as a matter of enlightened self-interest or as part of the socialisation into a way of life, a norm or rule exists as an inter-individual expectation which can straightforwardly be assessed in the usual decisionistic terms.

On this account therefore actions are not to be understood as produced by rules or norms but by actors' decisions to subscribe to the rule on some particular occasion. Neither the Winchian view of rules as constitutive of a form of life nor Durkheim's conception of behaviour being prescribed by norms is acceptable on this interpretation of conventions. Actors can always in principle opt out of a rule, it is just that in many instances the desire to avoid the sanction means that people conform more or less automatically. On other occasions rules have ceased to be regarded as external constraints and have attained an unquestioned status. In this state rules are usually referred to as norms. However, even here they are not finally compelling; the rational decision-maker view would interpret norms as forming part of an actor's wants or beliefs. If a norm against informing has lost the status of a convention and become virtually unquestioned it is still possible in a grave situation to feel that it is not worth submitting to an extreme penalty for the sake of the rule, perhaps because others will in any case inform. In the final analysis any rule or norm, no matter how deeply internalised, can be recognised as an invention and treated instrumentally and strategically. This is the essence of *Realpolitik*: any agreement, no matter how bindingly intended, can be treated as a cover for the pursuance of self-interest. No rule is beyond suspension. Logically therefore no rule determines action, only people's subscription to rules - a revocable decision - governs behaviour.

A few regularities appear not to have this exploitable character and seem not to be followed by choice but out of a form of necessity. Thus it might be accepted that rules for the use of symbolic systems, like conventional greetings or Durkheim's examples of currency and language, are not entered into revocably. Whereas the threat of sanctions or belief in the worth of a convict's code leads people to adopt it we do not have the same choice with language. As was discussed in Chapter 1, language appears to have a peculiar arbitrariness and can be understood purely as a device in the way that few other symbolic, structural systems can. It does not determine what people say only how it can be said and thus does not confute the thesis that rules are subject to a rational usage. Rules are the subject of behaviour, they do not determine it.

The theory of rational action allows that most rules may be followed on a great many occasions (although some exceptional rules may seldom be obeyed), but it crucially insists that people decide whether to obey rules and when greater wants intervene may ignore rules altogether. In this way rules and sanctions occupy a place alongside other of the agent's beliefs and wants. In some social scientific work, primarily in economics, it is supposed that wants are largely universal and that rules occupy a subordinate place. But this is by no means invariably the case: people's actions must be understood as a contextually generated response to particular wants and rules.

The traditional alternative to the analysis in terms of rational assessment depends on the concept of norms. Actors are understood as behaving in accordance with regular patterns which, in turn, are seen as dictated by societal demands. Unless norms are viewed as coincidental and random productions of history it is customary to understand them as fulfilling necessary functions. This whole issue is intimately connected to that form of sociological analysis known as functionalism which has been widely and justifiably discredited. The criticisms of functional explanations are well known and I shall not repeat them here since they bear no special relationship to the new philosophy of science.[15] According to functionalism, norms are understood once their contribution to the maintenance of social trends is understood. The flaw of course is that no agency is present to generate the norms in accordance with society's needs, nor is there an entity 'society' which can reasonably be said to have needs. What is of present interest about functionalism, however, is its method: using the terminology introduced earlier it is clear that the regulatory principle of functionalism is to inspect social institutions for their contribution to society's needs. Significantly, this regulatory principle depends on the analyst's ability to see a rational connection between an imaginary actor representing society and the creation of the prevailing norms as rules. Thus, the rational decision theorist can take heart that even functionalism appears implicitly to share a model of rational action.

When seeking to account for deeply sedimented norms there may be a temptation for rational decision models to slip into a form of functionalism, and the role of an imputation of rationality in both their regulatory principles can only make this possibility likelier. For example, in analysing capitalist society Marxist theorists are sometimes inclined to interpret all social institutions as beneficial to the interests of capital. Thus education is seen as establishing the kinds of discipline and attitudinal structure required in a capitalist workforce, and racialism is viewed as a means for depressing wage levels by dividing the workforce into strata of unequal pay. On a theory that power is concentrated in the hands of capital it makes sense to have a regulatory principle to inspect every social institution for its value to capital. But unless this regulatory principle can be limited, every institution will be held as an inexplicable anomaly until a class basis has been found for it. The method threatens that the theory will be forever self-confirming and effectively functionalist. Such a view paradoxically also threatens to make the possibility of working-class political progress undetectable since any apparent advances will be seen as covertly advantageous to capital. In a similar way any social theory which seeks to understand what (approximately following Hegel) may be called the cunning of history – the apparent co-ordination of historical events at a supra-individual level – is also confronted with a danger of lapsing into a methodological functionalism.[16]

Unless one attends to historical contexts the rational decision theory may suggest that actors' wants and beliefs are individual, historically unsystematic attributes. No matter how accurate and fruitful the economist's assumption that individual firms pursue profit it is important to acknowledge that these wants are foisted onto actors by the nature of competitive industrialised society. Rational decision theorists must accept that individuals with wants, beliefs and contractual rules tumble as it were through history. Indeed, the compatibility of this model of man with the Lakatosian scientist suggests that beliefs and rules of all sorts proceed in the episodic, non-linear manner of scientific research programmes. This progression through time is not unlike the natural selection of biological organisms but human rationality ensures that beliefs can be assessed by criteria which transcend mere historical fortune. The progression of wants is something on which rational decision theory is rather silent.

Beliefs, meanings and alien cultures

Since beliefs play such an important part in the rational decision theory it is worth considering the two principal problems in the philosophy of social science associated with beliefs. The first is generated by the private nature of beliefs, the second by the existence of bizarre or alien beliefs. The point about privacy has already been mentioned and can be dealt with quite briefly. Although unobservable, beliefs can be publicly scrutinised: if someone claims to believe something whilst acting in an incompatible way, such as believing that sunlight causes skin cancer whilst going on holiday to Tunisia, we do not have to accept that person's claimed belief. We would doubt that they really believed it unless it could be shown that they went to Tunisia out of some other interest and took precautions against the sun. The rational interpretation of action also supposes that there are canons of rationality against which people's claimed beliefs can be tested. The possibility is even allowed that people deceive themselves about their real beliefs.

A comparable analysis can be given of meaning: we can not mean just anything we choose by the words we speak or the acts we perform. Meanings too occupy a public realm which allows them to be scrutinised by other people. And this is not true only of what may be termed objective meanings, like words and symbols, but applies equally to subjective meanings. When someone receives news of promotion it would be expected to generate happiness and happiness would be publicly accepted. However the recipient of the news could claim to be disappointed and despite other people's protestations would insist it was bad news. Since the subjective meaning can be at variance with the objective meaning one might think that this demonstrates the untestability of meanings. On closer analysis the news could be understood as objectively disappointing if, say, promotion was a way of shifting someone from a job they loved. If no reasons qualifying the disappointment were provided we would not accept that the person was justifiably disappointed and would look for subconscious grounds for ennui. People do not have complete control over the depiction of their feelings or beliefs. These attributes can be treated as the object of scientific knowledge. Neither people's explanations nor even descriptions of their actions can be taken as unquestionable.

This complex of issues has often been seen to come to a head in the case of alien beliefs, for example beliefs in witchcraft or spirits. Such beliefs have often attracted scorn for their ignorance or they have been seen as akin to poetry, not real life beliefs but metaphors for undisclosed beliefs. This treatment has been promoted by the strangeness of the beliefs – beliefs so curious that they could not be scientifically motivated.

But such a judgement is only compatible with a philosophy of science which gives priority to facts. On a Lakatosian view a belief in spirits is not intrisically non-scientific. Indeed if Newton could scientifically accept alchemy then tribespeople can rationally believe in magic. The oddness of the beliefs is no basis for a test since some of the postulates of modern physics (massless particles and the like) are pretty bizarre. Provided only that beliefs are part of a research programme which is tested at the periphery they are virtually scientific. Tribespeople's potions may not work all the time but these instances can still be treated as anomalies in the absence of a better research programme. Lakatosians would argue that alien beliefs are not usually open to competition from other research programmes since the hierarchical society prevents other theories being introduced to rival the knowledge of witch-doctors. Their cognitive system is not unscientific but social institutions prevent the competitive assessment of alternative schemes. This state is not peculiar to tribespeople since in everyday life Western people may be similarly lacking in punctiliousness about their beliefs.

With culturally alien societies the practical problem is magnified since there are difficulties of translation even before their beliefs can be analysed. Someone who appears to claim that twins are birds may have a peculiar belief or they may have merely been misunderstood.[17] In essence the problem is the same as that of subjective beliefs: we would try to find other aspects of behaviour which relate to this belief, concerning, say, their treatment of young twins and young birds and so on. We impute both rationality and comprehensibility simultaneously. If we found a tribe who appeared to say one thing and do its opposite we would not accept that here was a group of irrational people, rather we would say we had misunderstood them. In this way the imputation of elementary rationality, in line with rational decision theory, seems to be the very precondition of anthropological fieldwork. Rational decision theorists regard this as another piece of evidence for the ubiquity of their theory. Equally, it makes them rather suspicious of authors who claim that the belief systems of other cultures offer some form of 'alternative rationality' – often a rationality which is implied to be superior to a selfish Western thought system. Yet precisely because the imputation of (our) rationality is the condition for making progress with translations of other cultures we could never discover an alternative rationality. Significantly, this is strictly analogous to the ethnomethodologists' claim to have found an alternative method for everyday thought. The new philosophy of science claims to have established the elements of rational thought *per se*. Both alternative rationalities and ethno-methods can be accommodated in the Lakatosian analysis of rational thought.[18]

Conclusion

The new philosophy of science does appear to make good many of the shortcomings of H-D naturalism and gives a more satisfying account of meaning, reasons and belief than do the arguments of anti-naturalists. Thus the new view finds a place for cognitive strategies reported by ethnomethodology without having to make everyday thought radically different from science, and shows how actions can be directed by reasons without letting reasons slip into unexaminable privacy. In this way reasons are shown to be a scientific explanation of action. Moreover, it offers guidelines of method in terms of developing research programmes. And although the philosophy does not dictate which substantive theories social scientists should adopt it does directly and indirectly endorse a theory of actors as rational decision makers. The regulatory principle following from this is to examine every action for the beliefs and desires which could have generated it. This principle holds good both for home and alien cultures. Only reluctantly should actions be regarded as irrationally brought about and only then when some obstacle to rationality – such as drunkenness or illness – can reasonably be imputed.

The history of institutional and supra-individual phenomena, including rules, must be understood as the production of (generally) rational actors and the regulatory principle should not be applied to society as such. In the course of the history generated by actors, beliefs and wants may change; although belief can be rescued from the fluctuations of history by the adoption of a strict Lakatosian methodology. While there are still difficulties in this theory – both with the theory of science (see the three points listed on p. 43) and with the theory of rational man (for example, the significant exceptions like mental illness are not yet understood) – these are sufferable anomalies. The theory is rationally and scientifically accepted because there is no superior research programme. The only alternative in sight, a neuro-physiological one, is massively underdeveloped. For these reasons the new philosophy of science looks with confidence on a naturalistic social science. In fact, that social science is already underway, it merely needs to become more disciplined and less side-tracked either by attempts to live up to a false image of science or by trying to find alternative rationalities and methodologies. Anti-naturalism is not proscribed – accounts of the operation of language or art appreciation are legitimate exercises. Such disciplines do not, however, tell us how social systems work nor are they the correct province of accounts of meaning and belief.

CHAPTER 3

Studies in the practice of science

Introduction

In Chapter 2 it was shown that the new philosophy of science was characterised by a concern with the scientific method. Scientific rationality was enshrined not in particular beliefs but in methods of assessing beliefs. This interest in philosophic portrayals of scientific procedure easily spills over into a consideration of scientists' conduct. The philosophy of science thereby becomes a theory of scientific behaviour. It was in this spirit, for example, that Popper's claim, that the essence of science is falsification, was denied; Popper's idea was faulty because scientists did not abandon falsified theories, instead they regarded falsifications as anomalies.

Given this dependence on claims about scientists' conduct one might reasonably expect that philosophers would have accumulated a stock of studies of scientific practice. But although the work of Lakatos and his followers is frequently built around illustrative case studies in the history of science the major compilation of empirical material has been undertaken by historians and sociologists of science.[1] To this extent the philosophy of science has maintained the philosopher's procedure (evident in ethics and morals) of examining abstracted accounts of behaviour. Even without philosophers' backing a large fund of empirical studies of science has developed in the last fifteen years. Although many of these studies have been encouraged by the absence of an orthodoxy as H-D interpretations of science fell, they have not precisely dove-tailed with the new philosophy of science. Rather than an attempt to reconstruct an epistemological theory of scientific behaviour they have been concerned with the personal and social pressures and influences which have led to the development of science at various times. In some cases this has generated claims that science is nothing like as systematic as the new philosophy seems to indicate.[2] On the other hand, historical studies have often not detected the massive problems of communication between research programmes that philosophers have anticipated.[3] Precisely because scientists may be working on a number of problems across various research programmes they are not so culture bound as abstract models suggest.

These empirical studies are not necessarily more nor less radical than Lakatosian theory would anticipate. They are, however, chiefly concerned with matters other than epistemology. On account of this non-epistemological focus such studies of science have sometimes been dubbed 'naturalistic', this does not imply that they are naturalistic in the particular sense being explored in this book. Some prominent studies of science claim to draw guidance from Winch and would thus appear to be anti-naturalistic.[4] Four of the major approaches to empirical studies of science will be reviewed in this chapter. The distinctive characteristics of each will be described and their implications for a post-empiricist understanding of science and of social science will be elicited.

Science policy, the sociology of knowledge and the political economy of science

Up to this point the nature of scientific knowledge and its implications for knowledge of the social world have been examined in principle. There has been an implicit model of science as concerned with 'pure' knowledge pursued by scientists whose only obligation has been to the subject. Crucially, no consideration has been given to the number of scientists available to work on any research programme nor on the availability of equipment. Neither has the model differentiated between research on, say, nuclear energy and archaeology in terms of their importance beyond 'pure' scientific considerations. In this section the relevance of this implicit model will be analysed to see how the questions of principle are affected by non-epistemological concerns.

Rather than begin with the contestable case of a 'pure' science, the issues can advantageously be spelled out with respect to applied science and its technological use. It is a commonplace in contemporary society that not every invention or potential innovation is taken up. Indeed economists typically distinguish inventions from innovations, the latter referring to inventions successfully adopted. Accordingly, one aspect of studies in the sociology of applied science is to explain why inventions are taken up and used at certain times and why they are brought to bear in certain areas. Why, for example, are cold-water detergents not generally produced in Britain and why are robots used in car production for the British market but less so for clothing manufacture? In general terms the answer is provided by an approach which may be termed the political economy of technology.[5] Fundamentally technologies are introduced when they are economically and, to a lesser extent, politically viable. Out of the pool of possible innovations only the economic ones are selected and this may well mean that research is directed towards inventions which are anticipated to be economically beneficial. Thus it might be the case that companies which are centred on some product (perhaps oil) would not welcome the introduction of techniques which drastically reduce their ability to make profits. Patents may be used to keep innovations off the market as well as to encourage invention.

In this way, because invention is closely co-ordinated with production the sponsorship of research and the overall dynamics of the development of techniques will relate closely to commercial considerations. Work on scientifically interesting and even socially beneficial inventions will find less favour than work on profitable innovations, and talent will be directed towards areas of achievement and high reward. The structure of research work and the type of inventions generated will reflect less about what people could make than about what the market would profit from. The realm of possible achievements may well be subordinated to the requirements of commerce. And this system may have grave and unanticipated consequences. It could be suggested that at times of recession, when innovation is needed, no private concern will have the resources to fund it. Equally, a fiercely competitive market may force profits down so much that there is no support for research. In a market economy the resources of knowledge are under control in the same manner as any other commercial resource.

Turning to academic science a similar kind of analysis can be put forward. Admittedly such research is not so immediately related to industrial and commercial needs - although much research is carried on in companies' R. & D. departments and at governmental defence establishments - and is often sponsored through disinterested bodies like research councils. Nevertheless, governments provide a sum of money which is not large enough to give all researchers the opportunities they claim to need so the logic of science alone can hardly be expected to determine how sciences grow when they are competing for livelihood. Moreover, market pressures of a different kind operate internally to science. Young researchers may well favour projects which can be completed quickly and which offer the scope for many publications because of pressures in the employment market. Such pressures, which can be termed demographic as opposed to nakedly commercial, must also inhibit the pursuit of any ideal, impersonal method.[6] It could, for example, be said to increase the amount of normal science and disincline people to develop revolutionary strategies. It has even been suggested that normal science can become exhausted so that there are no internal problems extant. In such a condition a research programme, devoid of anomalies and of new empirical inputs, may enter a stage of 'finalisation'. With all internal goals fulfilled the research programme can be harnessed to the solution of externally-posed problems. Thus as a crude example, the laws of motion may be understood and all that remains is the solution of problems about the trajectory of particular missiles or space vehicles.[7]

One rather speculative but interesting case study will serve to illustrate the political economy of science. In a study of the state of chemical knowledge Slack has suggested that modern chemistry is almost exclusively concerned with fast-acting and often irreversible reactions.[8] This he attributes both to the ready detection and to the usefulness of such reactions. Attention may have been directed to them in a disinterested way but this was reinforced by their assimilation into the chemicals industry which focused effort and funds into related research. Problems like the origin of organic molecules or the reversibility of reactions depleting our stocks of natural resources have been overlooked because they do not fit into this predominant framework of chemical investigation. Without necessarily endorsing Slack's claims it can be stated that by mechanisms such as those he proposed, forms of scientific knowledge can become adapted to the demands of society. A society or, perhaps better, certain relations of production can generate a state of knowledge which is compatible with economic needs (at least in the short run - one might add that the state of chemical knowledge has encouraged problems of pollution which are disadvantageous in the long run). Thus a Lakatosian logic of science would be a poor predictor of the actual development of science. The political economy of science could promote certain research programmes whilst denying funding to others. In many areas of science there can hardly be the kind of straight competition which Lakatos' view advocates.

This form of analysis is particularly attractive for Marxist writers who argue that the political economy of modern science will favour the interests of capital. Science can thus be said to be an ideological force. It is, however, important to be precise about the sense in which science may be ideological: the findings of science are not as it were falsified to restrain a proletarian science. Instead the choice of research topics and, to a lesser extent, research programmes is influenced by the demands of profit. The only strict ideological falsehood arises if it is claimed that science is autonomous and could not be harnessed to other goals. The theory of science, as much as the scientific theories themselves, is likely to be ideological. It may not be in the generalised interest that attention is given to heart transplants and sporting injuries rather than industrial illnesses, and this derogation from the general interest may be due to the political economy of science. This does not of itself make the theories surrounding organ transplant unscientific – though it may make the claim that nothing could be done about industrial ill-health untrue.[9] This viewpoint is not restricted to Marxist theorists. A pluralist could equally claim that the conflicting influence of power groups leads to the particular scientific theories we have at any particular time and that powerless groups are unlikely to do well out of this knowledge. In so far as scientists themselves form a pressure group they may opt for open competition between research programmes but if science is riven by personal and disciplinary conflicts then even an autonomous science could not hope to proceed in a Lakatosian way. The same holds true for governmental funding and planning of science. This analysis of the political economy of science is represented very schematically in Figure 1. Signals from the material world could in principle generate knowledge of many regions (although only a small number are illustrated in the

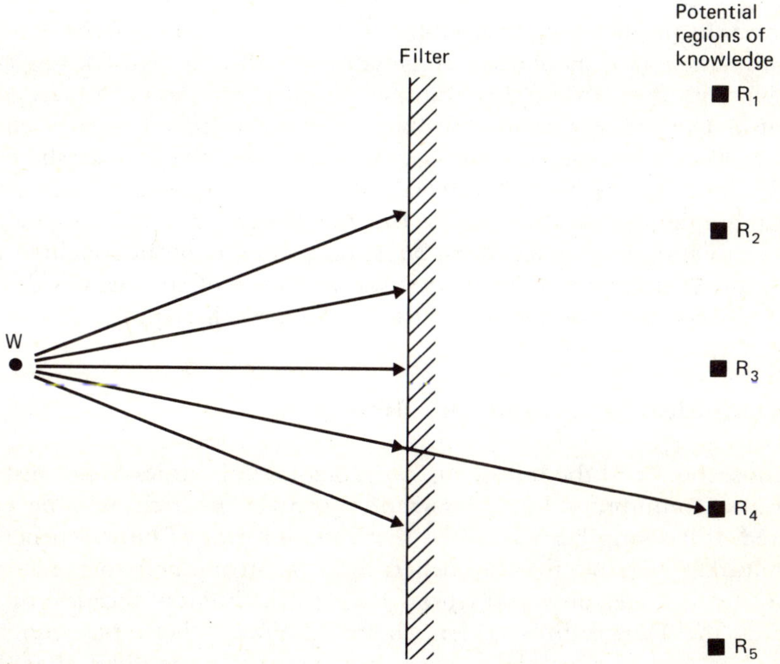

Figure 1 Illustration of the political economic effects on the development of scientific knowledge. Signals from the material world (W) which could lead to knowledge in a variety of areas (R_1 to R_5) are filtered by political economic processes so that only knowledge of R_4 develops.

Figure). At any particular time knowledge of some regions will be fostered while familiarity with others is prevented.

In so far as this analysis holds for natural science it must apply to social science too. It may even be more significant since the interests which filter natural scientific knowledge may have a deeper connection with the claims of social science. Thus a chemical monopoly which discourages scientific research on slow organic reactions through lack of relevance is judging the science on extrinsic grounds. It might have stronger reasons for discouraging research on work satisfaction and wage rates in the chemical industry. An analysis of the political economy of the academic establishment could similarly have much to say about the oft-remarked exclusion of women from the terms of reference of social scientific research, for example where families are categorised according to the social position of the father.

While the consequences of political economic factors may be very profound their direct effect on the question of naturalism is far less straightforward. The analysis of natural science in terms of political economy does not provide a favoured alternative to the Lakatosian method. Its implication, no doubt, is that science requires some form of co-ordination (and perhaps a socialisation of the means of knowledge production) but within this framework scientists should adhere to a scientific method. Practically speaking, both scientists and social scientists may well benefit from examining the selection of their research topics in the light of the fact that there may be unseen political economic factors impinging on their work. But this is equally true of naturalistic and anti-naturalistic endeavours. Most positively, one can conclude that the political pressures under which social science often operates do not of themselves make a scientific, naturalistic approach inappropriate.

The empirical programme of relativism

In this section, and the following two, sociological studies of not just the scientific community but of scientific knowledge itself will be considered. It has long been accepted that certain forms of human belief are so culturally variable that the beliefs must be attributable to the culture rather than to any universal truths. Perhaps the boldest assertion of this kind is the Durkheimian thesis about religious beliefs: that religious formations reflect the society's social organisation. A sociological explanation is sought for the contents of the (religious) knowledge.[10] So long as scientific knowledge was regarded as dictated by facts such a sociological explanation was regarded as totally inappropriate. The relaxation of the philosophy of science has been seized as an opportunity for sociological investigations of the content of scientific knowledge. The empirical programme of relativism (known as EPOR) is one of the dominant approaches to this task.[11]

This programme can conveniently be said to consist of three stages:

1 Revealing the openness of scientific results;
2 Examining the social processes scientists use in closing debates over results;
3 Studying the relation of these processes to social forces beyond the immediate social community of scientists.

Much of the work on the first stage of this programme was undertaken by Collins and the claims of EPOR can most readily be presented through a consideration of some of his empirical studies. Examining communities of scientists working with sophisticated, complex equipment he asked how directly their experiments led to the development of certified knowledge. He found that in areas of uncertainty and dispute unambiguous results just seemed not to be available. If someone tried to repeat another scientist's experiment - seemingly a standard way of checking results - results were never precisely repeated. Consequently sympathetic scientists would see the results as close enough and would explain away differences as due to slight variations between the apparatus. Hostile scientists would interpret the small differences as sufficiently large to invalidate the replication. Worse still, if exactly the same results were obtained this could easily be interpreted as extremely suspicious, indicating a massaging of results. Outside of textbook examples, scientific work is always so uncertain and so complex that the results cannot be persuasive of themselves. It is in this sense that scientific findings are open.

One of Collins' examples concerns the pursuit of 'gravity waves', a form of radiation thought to be generated by massive objects.[12] These waves had been predicted by Einsteinian theory but nobody had previously built an appropriate detector. Detectors had to be hugely heavy but simultaneously very sensitive to extremely weak forces. A detector could be set off as easily by a minor disturbance in the laboratory as by real gravity waves. On the other hand a machine built as a detector might have a sensitivity threshold just too high to detect the waves at all. In this way claims to have detected the waves could readily be interpreted in one camp as a mis-reading (possibly due to changes in the room temperature or something) encouraged by a zealous belief in the waves' existence. Pro-wave scientists, on the contrary, could treat repeated failures as due to insensitive detectors. Collins' claim is that no practically ascertainable evidence could settle this dispute. Furthermore, he argues that this is a representative instance and that no front-line, unresolved scientific issue could be adjudicated by the evidence alone since each side would be able to introduce sufficiently many *ad hoc*, but plausible, interpretations of particular results.

Now, this may appear only to be a reiteration of the point made in Chapter 2 about falsifications. Any result can be made compatible with a research programme through amendments to the protective belt. Lakatos maintains that a convergence can be generated out of this position in the long run. Collins does not accept this convergence: at one level he does not agree with it as a practical claim. Thus, because of the sheer number of differences between gravity wave detectors and the variety of assumptions which could be modified there is no single route to convergence. In this way contingent factors prevent Lakatos' scheme from working: scientific systems are too complex and, perhaps, aspects of the scientific profession discourage people from carrying out replications in a precise way. Collins suggests that gravity wave detectors may differ because there is less credit to be obtained from repeating something than from carrying out a new experiment. Under these circumstances successive gravity wave machines will differ even more widely than the complexity of the phenomenon dictates. Even without such contingent factors, however, Collins would not anticipate a workable Lakatosian convergence for he additionally suggests that scientists simply see the world in different ways and have unexplicated skills and assumptions which lead them to adhere to some particular point of view. Even if they wished to behave in a Lakatosian manner, in practice scientists could not. Working scientists are therefore faced with an irremediable openness.

According to Collins' research these factors generating openness are compounded by another attribute of modern science. Any particular research area may well be fronted by very few people and they, in turn, may be attached to a small number of expensive sets of equipment. These small groups effectively decide some area of science for the rest of the scientific community. Collins refers to these groups as core sets. Although small, these core sets usually comprise scientists working at different sites and with varying apparatus so that conflicts of interpretation can readily arise. Conceivably, a core set could form around one location or technique and establish a monopoly on the knowledge production. In such a case lengthy disputes would not be anticipated although it would be wrong to suppose that the absence of dispute meant that the knowledge was uniquely correct. In most cases however the situation is oligopolistic. It is hard to imagine a state of affairs more antithetical to a Popperian or Lakatosian view of science as a cognitive free market. Because scientific knowledge lies in the hands of such a small number of people who are faced with great interpretative openness, Collins considers that a stance of relativism is profitable in studying scientific knowledge. This need not be an extreme relativism (suggesting that there is no fixed real world); it merely proposes that the contents of the world are irrelevant in explaining what results these small groups finally settle on. Again this claim is somewhat ambiguous between the assertion that for practical reasons the real world cannot get through to people working within professional and temporal constraints and the stronger assertion that scientists' tacitly acquired skills and predilections dispose them to see the world in different ways – incorrigibly.

This postulate of relativism is both a methodological guide for sociologists (almost a regulatory principle) and an expression of the state scientists actually find themselves in. Faced with such complexity and diversity scientists could not provide an unambiguous depiction of the natural world even if their career patterns encouraged them to try. Yet as an empirical matter scientific disputes seem to wax and wane and the histories in scientific textbooks give accounts of decisive evidence and critical tests which allegedly resolved past disagreements. The relativist is unmoved by such accounts which have to be regarded as a form of rationalisation. Some social mechanisms must be responsible for the achievement of the closure which the world cannot perform. Even in those historical instances which the textbooks record, at the time of closure the evidence could not have been conclusive (according to EPOR) so the agreement must have been socially achieved.

This second stage of EPOR has received less attention than the first but a brief statement of two possible mechanisms will indicate what Collins has in mind. A mechanism relatively peculiar to the scientific community is the use of measurement.[13] Collins has suggested that measurement provides a ready and superficially compelling procedure for deciding opposing claims. Instead of saying either the results are very close so the detector must be working or the results may look close but they are not close enough, measuring instruments allow you to say how close. A machine can be calibrated and then used to make a series of impersonal tests. However, it can readily be argued that the machine merely obscures the problem rather than solving it. It remains to be argued over whether readings which, say, vary by five per cent are to count as similar or different, and of course any machine can become a focus of argument: it is reliable? does it really measure the things we think it does? and so on. None the less, if consent can be obtained for the use of some machine then arguments against the machine begin to look like special pleading. There might well be an advantage, therefore, in being on the side which introduces the measuring machine since it can embody your point of view. A fanciful example is the kind of calibration that heavy drinkers might have used if they had been the sponsors of breathalyser devices.

A second kind of mechanism can be displayed through Collins and Punch's study of parapsychology, a study which will also introduce stage three of EPOR.[14] Collins and Pinch made an extensive study of the arguments over parapsychological phenomena. At one level this permitted them to show the theory-ladenness of both sides' observations: where some people see psychic bending, others see surreptitious cheating; where heat is detected some interpret it as atomic reorganisation under parapsychological forces, others regard it as friction generated in cheating. Supporters of both sides have aimed for a definitive proof, through successively more elaborate tests, but opponents are claimed to remain stolidly unmoved. Collins and Pinch record an absence of any evidence which is agreed to by both sets of protagonists. Hence no Lakatosian comparison can be undertaken. There are, however, changes in the fortune of parapsychology over time, Collins and Pinch regard it as having become more established, and these changes can only be accomplished through closure mechanisms. The authors see a strategy in terms of activities like founding parapsychology journals in a scientific style, obtaining research council grants for study of parapsychological phenomena and affiliating to national academies of science as the key to parapsychology's success. Once these steps have been taken it becomes extremely difficult to discount parapsychology altogether and much easier to entertain the subject as credible.

Since these aspects of the latter closure mechanism may depend on wooing publishers and scientific academies it is plain that some closures depend on politics in a wider sense than the negotiations over measuring devices. An academy which was suffering economic stringencies would possibly be ill-disposed to the acceptance of a new and risky 'science' of parapsychology whereas a scientific establishment being pilloried for an over-conservative approach to acupuncture might welcome another radical form of science. Other conceivable third stage scenarios will no doubt come to the reader's imagination.

As has already been mentioned with respect to Kuhn and Feyerabend relativism is antithetical to the mainstream philosophy of science. EPOR might therefore be dismissed out of hand unless the particular sense of relativism it involves is understood. Collins is not committed to saying that people can believe just anything, far less is he claiming that scientists make the wrong choices. He is simply insisting that there are always insufficient cognitive grounds for reaching any conclusions and that various social mechanisms must eventually enforce the closure which cognitive factors cannot make. Since both parties to a dispute usually have quite elaborate arguments to support their case, whichever side achieves closure will have a pretty effective theory to be getting on with. Any scientific research programme is likely to offer many satisfying explanations.

A common retort to the claims of EPOR is that they are self-refuting. It is sometimes held to be self-contradictory to assert that relativism is true and EPOR is bracketed into this form of relativism. But it is not obvious that it belongs there at all: briefly expressed, Collins need not claim that EPOR is unquestionably correct, only that it is a reasonable research programme. Accordingly, his support for EPOR is no more, but equally no less, thorough than any scientists' belief in his or her theories of chemistry or whatever. Furthermore, since natural scientists commonly behave as though knowledge was not open Collins argues that this form of 'naive' empiricism is a customary attitude of research (rather like the unquestioning attitude of normal science) and an attitude which can reasonably be adopted by practitioners of EPOR. The only component lacking is an explanation of why this attitude should be both pervasive and effective. That it is present can be attested by the repeated attempts by parapsychologists to do a definitive experiment and the uncomprehending disappointment when the experiment 'succeeds' but no-one is converted.[15]

By its very nature EPOR does not lend itself to the provision of methodic guidelines for research in the same way as Lakatos' analysis. On the other hand it does offer strong practical advice to expend energy on the public profile of social science. It suggests that neither the merit of particular theories nor of social science as a whole can be practically established through cognitive activity alone so the best strategy must be to be tirelessly evangelical about one's theories. With respect to the naturalism/anti-naturalism issue, Collins' strictures would apply to cognitive ventures of either kind. It is noteworthy, however, that even though EPOR is often traced to Winch's point of view, the actual description of the closure mechanism and research strategies is thoroughly naturalistic. An open, quasi-relativistic natural science can still offer a workable model for social scientific investigation. In a curious way, therefore, EPOR endorses Lakatosian recommendations since no better method for scientific investigation is offered than the idealised vision of competing research programmes. According to EPOR, scientists get on as well as they do through a form of naivety. EPOR offers no reason why the majority of social scientists should not also proceed in this way, confronting incorrigible openness with a belief in empiricism.

Social interests and scientific knowledge

Although EPOR has provided strong arguments for the practical inapplicability of method and for the pragmatic value of relativism a more explicit attempt to generate a social theory of scientific knowledge has appeared in the form of interest theory. This approach can be traced in general terms to the work of Habermas discussed in Chapter 1. It will be recalled that according to Habermas our ability to comprehend the world at all depends on selecting out certain information on the basis of characteristic human interests. Hence knowledge of the natural world supposedly relates to an interest in its instrumental exploitation. In Habermas' view all science was based on the same enduring interest; he considered interests to have a quasi-transcendental status. The approach I am terming 'interest theory' argues instead that scientists have various empirical, contingent interests and that these interests dispose them to interpret the world in differing ways.[16]

It was the starting point of many EPOR analyses that there were two alternative points of view (for example, pro- and anti-gravity waves). No matter how opposed to relativism one is, it is impossible to maintain that these alternative viewpoints could have both been brought about by the same evidence from the natural world alone. Interest theory would seek to explain why some scientists came to be pro-wavers while some derived an anti-wave interpretation. In principle the argument would be that one group's interests inclined them to the existence of waves while the other group's did not. In this way a fundamental but contingent association is sought between a social attribute (interests) and the content of knowledge. Interest theorists maintain that sides in a dispute form around interests and that the outcome of the dispute is decided through the social and sometimes political competition of these interests. It may be the case that interests relate straightforwardly to the knowledge claims associated with them (thus the rich might believe in the inevitability of wealth inequalities) but such an internal connection need not always be present. The knowledge/interest connection may be very indirect: hence traditional traders may favour non-metric weights not because there is anything radical about metrification per se, but because the process of metrification will create opportunities for members of other professions (the manufacturers of scales) to gain a trading advantage. Crucially the connection between knowledge and interests is always of an empirical nature.

The application of interest theory to the analysis of science can usefully be introduced through consideration of one of the clearest and most widely known case studies in interest theory. Around the beginning of this century two statisticians, Pearson and Yule, set out to provide a measure of association for nominal variables. The desired statistic would allow one to establish, for example, the degree of association between eye colour (itemised as blue, brown etc.) and right- and left-handedness, and to compare this with the association of eye and hair colour. According to MacKenzie's analysis the statisticians produced rival measures having rather different properties.[17] Thus, Yule's measure was easy to calculate whereas Pearson's was complex but closely integrated with established statistical measures of interval variables. A considerable dispute broke out between these two men over the worth of their measures. In this dispute, MacKenzie suggests, both men defended their own statistic as corresponding to the real degree of association and claimed that their opponent's statistic was false. Not only had these men produced different statistics, they also had different criteria for assessing the quality of a statistic. No Lakatosian assessment could get underway because the men could not agree on what constituted a progressive development of the research programme nor on which programme generated the most significant excess empirical content. Unlike the analyses of EPOR which revolve around the impracticability of the Lakatosian scheme under radical uncertainty, interest theory suggests that the same interests which lead to divergent research programmes will generate divergent criteria for the comparative assessment of those programmes. On this view the sides to a dispute approximate more to Kuhnian paradigms than research programmes since both the empirical claims and the criteria for assessing those claims are restricted to members of the paradigm.

As was also shown to be the case with EPOR, there is a similarity between interest theory and Lakatos' theory of research programmes. Interest theory acknowledges that on both sides of a dispute work goes into extending knowledge. In the case cited above both Yule and Pearson derived new additions to their measurement of association, partially in response to their opponent's criticisms. Opposing research programmes are not mere rationalisations of the interests they serve. Barnes, who has done most to conceptualise this approach to science, acknowledges that humans have a 'natural rationality' which is used to elaborate frameworks of knowledge.[18] Deep down, however, opposing research programmes embody and are sponsored by social interests. In Pearson's case, it is suggested, his desire to measure intelligence (intelligence was then assessed as high and low) for eugenic reasons led him to want to develop a statistic compatible with pre-existing analyses of heredity. In its turn, his commitment to eugenics was associated with his class position as a member of the upwardly mobile, professional middle classes. Contingent social interests, whether they are of this direct class type or much more temporary, as in a desire to succeed in a chosen branch of chemistry, constitute the basic perspective around which a hard core of belief forms. Conflicting research programmes represent conflicting interests and although natural rationality functions in the normal procedures of articulating the research programme, only a closure accomplished by the triumph of one interest can bring a dispute to resolution. As in EPOR a dispute can only be closed by social mechanisms, but interest theory diagnoses the openness differently from EPOR. In the latter openness results from practical ambiguity, in the former it results from systematic differences of perception.

Interest theory has benefited greatly from its ability to generate empirical case studies. Yet impressive though they are, they do not remove a possible doubt about the general validity of the theory. Even if a great number of case studies had shown that alternative interpretations of scientific evidence were predicated on divergent interests and that the knowledge which was accepted in the long run was that associated with the dominant social interest this would not prove that disputes could not be settled disinterestedly. In the first place a Lakatosian could accept that alternative viewpoints were generated by different interests: for example, people who benefit from free trade could be expected to build a theory around the value of free trade whereas those who suffer from it could be expected to pioneer a theory of the inadequacy of free trade. Just because these beliefs reflect people's interests that does not prove that the wealthy could not be persuaded that they would be just as well off in a planned market or the poorer people persuaded that without free trade they would be poorer still. Even if the rich prevented the theory from being properly tested so that their interest foreclosed the debate, that does not prove that no disinterested assessment could be made. No collection of case studies could demonstrate that disinterested assessment is impossible (although much like EPOR, they could show it was very improbable!).

To meet this problem theoretical analyses in interest theory have been very important. One strand in this theory is represented by Barnes' book *Interests and the Growth of Knowledge*. On this view alternative theories or research programmes are rival depictions of the world. Part of their contents is given by the input of previously certified experiments and so on but this input under-determines the theoretical structure which it supports. Hence the remaining determining influence is provided by socially generated criteria. These criteria may differ but they need not. And if a criterion for maximising our control over the natural world were introduced, following Habermas, then science could operate like a learning machine. On this interpretation, an interest in instrumentality becomes a specification of what we used, naively, to call disinterestedness.

In other formulations of the issue Barnes has departed from this interpretation and claimed that instrumentality is itself a flexible notion which will be understood differently by people with different interests.[19] This appears to endorse the position adopted in many of the case studies. However since this claim appears fundamentally similar to the EPOR assertion that scientists will not arrive at a practicable agreement, and does not provide any proof that agreement about the meaning of instrumentality could not be reached, there remains room for doubt about the claims of interest theory.

Accepting this cautious evaluation of the theory it is still possible to view it as confirming the claims of EPOR about the unlikelihood of effective Lakatosian resolution of any active scientific dispute since social closure mechanisms are widely available. Furthermore, it allows a greater elaboration of the political economic view of science as ideological. As

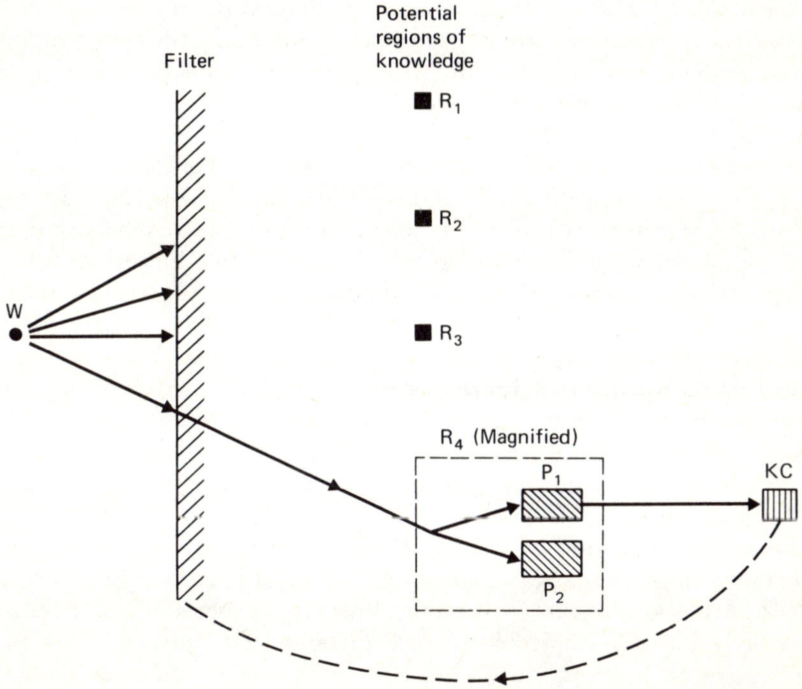

Figure 2 Illustration of the combined effects on the development of scientific knowledge of political economic factors and of social interests. Within the region (R_4) encouraged by political economic factors, two competing research programmes (P_1 and P_2) are sponsored by varying interests. The question of which knowledge claim (KC) wins acceptance is determined by the contest between the interest groups. In some cases the knowledge claim that wins acceptance may serve to reinforce the legitimacy of the filter – this is indicated by the dotted line.

alternative interests commonly sponsor the available research programmes social interests (perhaps of a class-based nature) will be operative both in selecting the regions to be studied and in formulating the hard core of the research programmes. And if, as often seems to be the case, closure is achieved through social mechanisms then interests may affect the content of established knowledge. This extension of the political economic analysis is represented in Figure 2.

As with EPOR this analysis holds as much for social science as for natural science and for anti-naturalistic as much as for naturalistic studies. The connection between the content of social scientific theories and the sponsoring interest may be more literal and direct than is usually the case for natural science. Whether this makes social science an easier vehicle for ideological rationalisations than natural science or whether it makes ideology in social science harder to conceal is a matter for debate. Interest theory offers no direct guidelines of method although it suggests that we cannot but benefit from analysing the reasons for adopting the central assumptions of our theories.

It is finally worth remarking that the model of the actor in interest theory is very similar to that of rational decision theory. People act in accordance with their interests and beliefs, beliefs which are derived from their interests by natural rationality. Of course in interest theory beliefs do not offer a dimension of the actor on which context-free judgments can be passed, but interest theoretic analyses are themselves resolutely naturalistic.

Practical reasoning in scientific work

A recent trend in science studies has been for analysts to examine scientific work and life in laboratories in considerable detail. Instead of seeking explanations for large-scale cognitive entities like the atomic theory, analysts have become interested in how agreement is reached over specific findings and how negotiations are conducted at the laboratory bench. Attention is directed at the everyday aspects of science, and scientific activity is depicted as social through and through - no more nor less so than work in an office or mine. Scientific knowledge is viewed as the product of negotiations and interactions between scientists. Epistemological concerns are side-stepped.

Within this empirical programme two forms of study have been emerging which have recently been dubbed studies *of* and studies *in* scientific work.[20] Studies of laboratories and scientific work include ethnographic and microsociological studies of how scientists generate scientific findings. By studies in scientific work is meant analyses of scientific reasoning which to a large extent are self-exemplifying so that the method of anlaysis itself becomes an object of study. The reasoning procedures of scientists and sociologists alike are subject to examination. Studies of scientific work have come to be termed 'the programme of constructivism', and it is with these that I shall begin.

A leading exponent of this approach, Knorr-Cetina, characterises it as demanding 'direct observation of the actual site of scientific work (frequently the scientific laboratory) in order to examine how objects of knowledge are constituted in science'.[21] Whilst this general ambition is not predicated on any specific theoretical view of science it is possible to outline the kind of analysis this proposal generates. Scientific knowledge is viewed very literally as constructed; it is cobbled together in the light of the resources and obligations of the laboratory in a way made possible by the skills and abilities of the scientific staff. Hence a vast range of pressures and goals must be balanced in the execution of scientific tasks. And this ensures that knowledge does not grow in any observably methodic manner. For example, scientists may be constrained by the kinds of reagents available to them and may spend much of their time getting apparatus and experimental rigs to work satisfactorily. They may have to decide between alternative measuring devices and, if an experiment produces a large number of results, they will have to select a number of 'representative' instances. Because of the unavoidability and multiplicity of these choices scientific work is said to be decision-impregnated.

Other factors also lead to scientific work being unmethodical and 'messy'. Everyday misfortunes may hamper work: there are many stories of power cuts bringing about the irremediable thawing-out of frozen samples. And this recalcitrance in the world of objects is magnified in studies of organic topics: animals may die unexpectedly or refuse to die neatly for the researcher's convenience. In all these respects scientific work is highly circumstantial and the knowledge generated cannot but have the quality of an artefact. A kind of pluralism frequently pervades these studies leading to claims that any number of personal aims and interests, together with negotiated compromises, lie behind the work done and the results produced. For this reason exponents of constructivism may well be wary of interest theory since interest theory effectively demands that people have stable, identifiable interests which can be harmonised with the knowledge they produce. Constructivism insists that knowledge is far more haphazardly generated although it is not thereby any purer than interest theory would suggest. At the level of methodic recommendations constructivism offers no alternative to the new philosophy of science except to say that any pursuit of pure methodicality is bound to be self-deceiving or worse. The pursuit of a procedure without using arbitrary decisions would lead to a regress preventing the original project from ever being undertaken.

In contrast to the optimism of constructivists in carrying out 'direct observation of ... scientific work' analysts concerned with studies in scientific work are unrelentingly attentive to their techniques of 'observation'. An understanding of science is sought as much through the process of analysis as through the material analysed. In turn this examination of analytic procedures has produced results of two kinds. The first kind consists of details of scientific reasoning. Woolgar, for example, has shown how scientists appear to be sociological as well as scientific reasoners. That is, he has found that in understanding each other's behaviour and assessing others' beliefs scientists take into account their colleagues' social positions.[22] Hence scientists constantly impute interests and beliefs to each other which create a reasonable context for an overall assessment of belief. Indeed, in practical situations scientists often appear not to discriminate between 'social' and 'scientific' factors in coming to evaluations. In a similar manner Latour and Woolgar have indicated how in the course of making a claim about the discovery of a new fact, substance or finding scientists change the status they ascribe to their results. What, initially, the scientists refer to as a product or artefact of laboratory procedure is subsequently ascribed a natural and impersonal status.[23] Hence, something which starts off being described as the outcome of a laboratory process later comes to be seen as the causal agency behind this process. In this way the local history of scientific work and of scientists' own thought is constantly being revised and updated. According to this view the procedures of scientific research could not be said to be directed by any single philosophy of science; scientists themselves appear to oscillate between conventionalist, realist and even sociological views of their own and others' work.

A second direction for analysis has been towards sociolinguistics and a study of scientific discourse. Advocates of this perspective argue that scientists portray the practice of science very differently in a variety of contexts. Significantly, scientists appear to have several linguistic registers for the description of science and these registers are regularly used in different contexts. For example, an impersonal empiricist presentation is commonly found in formal scientific papers whereas a far more idiosyncratic, messy depiction is provided in informal contexts. The analyst of science is therefore faced with a variety of accounts of the specific actions which occur in the course of scientific work. The obvious path to generating an orderly description is to select one account as basically sound and to discount the alternatives. However, analysts report that in the course of interviews with many scientists working on related topics accounts of such complexity and mutual incompatibility are produced that no single account can be taken as generally preferable.[24] Only an essentially arbitrary process of selection could reduce the accounts to order. Instead of adopting this option analysts have chosen to study in the first instance the constitution and nature of the various accounting registers.

One consequence of this decision to study regularities in accounts, or discourse, has been a critical appraisal of other trends in science studies. Generally speaking ethnographic analyses appear to give priority to informal accounts while philosophical studies, which commonly depend on the published literature of science, grant precedence to formal discourse. In neither case is explicit attention given to the accounts which have been screened off. This issue appeared in the discussion of EPOR too, where it was noted that EPOR offered no explanation of why scientists themselves seemed to adhere to a positivistic interpretation of science. The persistence of this accounting register was left unexamined.

The thrust of this kind of work is not, however, exclusively critical. Setting out from a critical inquiry sociolinguistics may lead to substantive findings. Thus science is often characterised, not least by Lakatos' followers, as built around the attainment of convergence and consensus whereas both EPOR and interest theory demand that there are irreconcilable 'sides' to the majority of scientific issues. An analysis of scientists' accounts on the other hand suggests that sometimes scientists will stress the unity of their community and sometimes they will portray it as divided. And these depictions seem to depend less on any attempt to guage the degree of consensus than on the demands of the immediate context in which the account is given. Strong claims for consensus are often provided when the speaker/author is evaluating a knowledge claim highly ('most scientists now accept that ...') while claims for widespread disagreement are associated with attacks on knowledge ('Jones thinks he has the correct interpretation but most people only go along with him for part of the way'). In the first instance at least, the interpretative significance of the ascription of consensus is as noteworthy as the question of whether consensus really exists in the scientific community.[25]

Perhaps the most striking successes of this perspective have been in studies of scientific culture slightly removed from the epistemological questions confronted by Lakatos, EPOR and interest theory. Thus, for example, the humour which is often reported in scientific work has received little attention from epistemological analysts while sociolinguistic studies have been able to suggest that scientific jokes are generated out of the incongruous conjunction of different registers of scientific discourse. The contrast between the impersonal formal register and informal accounts of the messiness of scientific practice provides the material for many humour remarks.[26]

As opposed to studies *of* scientific work which have many affiliations to EPOR, studies *in* scientific work represent an increased self-consciousness within studies of scientific practice. Studies of practical reasoning suggest that the imputation of social interests is just as much in need of analysis as the scientific work which interest theory set out to explain. Studies of scientific discourse argue that many philosophical and sociological analyses of science depend on unexamined methodological practices – unnoticed practices similar to those which fuelled the arguments of those sociologists who adopted social phenomenology and ethnomethodology in preference to former H-D models. In this way social studies of science have become less a theory of science than a model for the study of society. Studies *in* scientific work have moved from naturalistic studies of scientific practice to virtually anti-naturalistic explorations of scientific culture. There remain many problems with studies in scientific work, notably the rather vague dependence on the concept of context and the imprecise formulation of the registers of scientific accounting. But the persistence of these methodological problems in an area of sociology where, because of its close connections with the philosophy of science and scientific practice, an understanding of method could be expected to be unusually high indicates the continued significance of studies of method.

Conclusion

The philosophy of science associated with Lakatos depends on a theory of scientific behaviour. It accordingly invites either confirmation or opposition from alternative, naturalistic analyses of scientists' conduct. Contributions from political economy, EPOR, interest theory and studies of scientific work confronted Lakatos' work with ironic results. For a variety of reasons, ranging from the pragmatic to the principled, it was claimed that science could not be expected to proceed in a manner which Lakatos would accept as rational. However, this was not because some other rational approach was detected. Scientific development was social and largely non-rational. Yet when it comes to making recommendations for the method of social science the absence of an alternative theory of rationality means that, with pragmatic reservations, a form of Lakatosian method is recommended.[27] Moreover, since both Lakatos and these analysts of science are offering naturalistic theories of how scientists, as people, behave they are all open to anti-naturalist objections. It is to an examination of the practical upshot of this curious set of ironies that I must now turn.

CHAPTER 4

Post-empiricism revisited

Philosophy into practice

Up to this point the discussion has chiefly concerned which of the available theoretical interpretations of scientific knowledge is most appropriate to the natural sciences and what implications this holds for our understanding of social knowledge. Among philosophical accounts of science, Lakatosian post-empiricism has stood out as generally preferred and although it has been subject to great criticism from empirical accounts of scientific practice these latter accounts have not offered any alternative prescriptive interpretations of science. It is now time to assess the usefulness of this post-empiricist prescriptive theory for social scientific practice. This assessment can be seen as composed of three parts: whether the theory holds good for natural science; whether the account is actually beneficial in practice; and whether any peculiarities of social science throw its promised utility into doubt. Although the correspondence will not be exact, this section of the chapter will largely deal with the first of these issues, the next section with the second and the remaining sections with the third.

Turning first to the appropriateness of post-empiricism as an account of natural science, it has already been suggested in Chapter 2 that there are outstanding difficulties in the formulation of Lakatos' analysis. While many philosophers would no doubt accept the basic Lakatosian model of competing research programmes, a general agreement is no substitute for strictly specified rules of method. Precisely because Lakatos' theory of science centres on a decision procedure for evaluating research programmes (i.e. because it locates rationality in method) it demands exact formulation. As it stands the model is open to criticism as a decision-making algorithm. The model lays responsibility for evaluating competing programmes on the choices of individual scientists. However, in any scientific dispute it is inevitably the case that both programmes can be rationally supported in their own right since each will typically have predictive and explanatory strengths. They will, in short, both be *scientific* research programmes. A clear, albeit simplified example of this was provided by the competing theories of rock formation described in Chapter 2. Whereas the thermal interpretation easily handled the phenomena of insoluble, quartzy rocks, the chemical version suited minerals which dissolved in water and substances decomposed by heat. Lakatos' model suggests that the allegiance of competing scientists to their own research programmes will lead to the internal development and elaboration of the programmes until one or the other commands greater credibility. Even supposing that such pure competition could exist between alternative theories, this decision-making mechanism rapidly runs into technical problems.

The first difficulty is that Lakatos' theory does not specify when a scientist's allegiance should switch. There is no rule or method which determines when one programme has left another so far behind that the trailing programme should be deserted. Like business-men dropping an unprofitable line, scientists are supposed to have the sense to leave a failing programme but the very fact that scientists make their departure at different times indicates that there is no criterial point at which a programme is to be deemed hopeless. Seen in retrospect perhaps, scientists may have switched their commitments at about the same time but there is no single instant at which such a switch becomes rationally compelling.

This weakness could be avoided if, as has been ironically suggested, a time limit could be imposed.[1] Were it possible to say that, for example, if one research programme is continuously weaker than another for a year then it could be rationally discarded, then Lakatos' scheme could operate as an algorithm. But any such limit would clearly be arbitrary: some experiments or observations take great periods to set up and some aspects of animal evolution or geological development may be unnoticeable except over several lifetimes. Yet without either a time limit or some decisive criterion for theory rejection it becomes rationally defensible to commit oneself to any scientific research programme on the off-chance that it may come good again. With sufficient *ad hoc* adjustments Newtonianism could be made to accommodate the results of modern physics and it must remain forever possible that a revised Newtonianism could generate some unanticipated prediction. Thus, contrary to many people's intuitions, Lakatos' theory does not indicate that any research programme is beyond the rational pale. And just as it tolerates scientists who wish to revitalise discarded research programmes, it cannot instruct any scientist that it is rational to give up the programme on which they are currently working. Naturally, historical case studies written from a Lakatosian point of view claim that all reasonable scientists moved to the progressive programme at around the same time, usually under the influence of the same results. This category of 'reasonable' actors is, significantly narrower than the notion of 'rational' actors;[2] what tends to happen is that scientists who hang on to their former beliefs are gradually frozen out of funds, publications and research opportunities so that for pragmatic reasons they conform or simply disappear from the research community. Even while accepting Lakatos' general account therefore one can be very doubtful of its status as a precise guide for scientific activity. Indeed if pressed to its logical extreme the action Lakatos recommends could justify scientists in subscribing to a far wider range of beliefs and programmes than is found in practice. Far from directing scientists in what they should believe it offers them a wider range of possibilities then everyday experience would suggest. And if the scientists can rationally adopt any of a number of alternative theories it becomes easy for the sociologist of knowledge to argue that extra-scientific interests or influences actually determine which programmes scientists do adhere to.

It appears evident that this (sociological) conclusion is not one to which Lakatos is attracted but the fact that it can be derived so easily from his theory indicates very strongly the theory's lack of specificity. While his theory provides a plausible portrayal of the general considerations which enter into scientific assessments it fails to demonstrate that any evaluation is rationally compelling at any particular point. It aims to locate scientific rationality through the aggregation of individual scientists' reasonable decisions. But this dependence on aggregated assessments means that it is ill-equipped to direct individual evaluations.

Much of the support for the Lakatosian theory derives from historical studies where it is claimed to be shown that scientists' actions accorded with Lakatosian expectations. There is, though, something peculiar about this appeal to history. First of all there is the problem mentioned above that it is much easier to show that it appeared reasonable for scientists to switch their allegiance than it is to demonstrate that they adopted the only rational course. In any historical instance it is hardly surprising that scientists who changed their commitments had some good reason for doing so. It is then all too easy, especially with the benefit of hindsight, to identify these reasonable grounds with rational inevitability.[3] In this way virtually any historical incident can be viewed as a confirmation of the Lakatosian viewpoint. I cannot envisage any practical way of establishing that scientists' choices were made out of rational compulsion rather than out of a human judgement passed on the basis of reasonable, but by no means incontrovertible, grounds.

A second, more abstract issue concerns the principle of basing a philosophy of science on the aggregated history of scientists' choices. At one level this principle has the ironic consequence of making our knowledge of science depend on our knowledge of history. Hence our theory of science can only be as good as our grasp of history allows – an unexpected reversal of the priority usually accorded to natural science over social knowledge. Additionally, this dependency on history precludes the possibility of asking questions about the transcendental validity of science (such as perhaps Habermas or, more classically, Kant and Hume ask). Even if the history of science provides a succession of theories in Lakatosian order, by what criteria are we claiming that later theories are superior? In short, what makes it good or worth while to play the game of cognition according to Lakatos' conventionalist rules? Realists aim to meet both aspects of this historical problem by saying, respectively, that a true knowledge of history can yield a real understanding of science and that, in any case, successive research programmes provide ever more precise realist depictions of the universe.

No matter how attractive these realist assurances sound they cannot remove the difficulties because there remains the problem of proving that the Lakatosian method actually does subserve realism. There is no doubt a strong temptation to suppose it does, but if the subservience is to be demonstrated empirically, say through historical example, then the historical problems reassert themselves. On the other hand, any purely philosophical proofs of the subservience would have no use for the historical confirmations: if scientists had not behaved this way in the past so much the worse for them. By lashing itself to history and human conduct Lakatos' theory of scientific rationality denies itself the one road to certainty. Lakatosian scientific method remains a convention – perhaps a popular and credible convention – but a convention without ultimate justification.

The applicability of the philosophy of science

At first sight one might well expect that because post-empiricist philosophies of science concentrate on methods of scientific conduct they would be directly applicable. Indeed, Popper's *The Logic of Scientific Discovery*, one of the first philosophies of method, gives strong hints that it is intended as a guide to action.[4] As we have seen, Lakatos' work is most often 'applied' to historical instances. Its application is aimed at showing how past scientists acted rationally – that is that although they were not pure empiricists, they in fact acted in an optimally rational manner. In a sense therefore these applications serve to retain the dignity of past science, but is that their only aim? Should not another aim be to guide contemporary science, as with Popper? Lakatos seems to think not:[5]

> Methodological progress still lags behind instinctive scientific verdicts in the sense that the main problem is to find, if possible, a theory of rationality which would explain *actual* scientific rationality rather than to bring legislative interference by the philosophy of science to the most advanced sciences.
>
> *Secondly*, I hold that the philosophy of science is more of a guide to the historian of science than to the scientist. Since I think that philosophies of rationality lag behind scientific rationality even today, I find it difficult fully to share Popper's optimism that a better philosophy of science will be of *considerable* help to the scientist; although no doubt it may help – and Popper's philosophy *has* helped – those great scientists whose scientific judgment was warped by the influence of previous, worse philosophies.

Unsympathetically summarised this statement might be taken as asserting that the new philosophy of science can at best rescue science from some of the delusions previously foisted on it by earlier philosophers of science. Aware of philsophy's fallibility the humble philosopher might undertake to give up consorting with scientists altogether. Implicit within this statement there are, however, two positive claims. The first is that the specific errors induced by past philosophies may be corrected. In the case of natural science this might be said to apply to the case of operationalism, a philosophy introduced earlier this century to account for difficulties of measurement in physics. It was proposed that physical properties, like temperature or dimensions, should be expressed in terms of the operations used in recording them. Hence spurious disputes deriving from measuring the 'same' properties with different instruments would be eliminated. But the new philosophy of science is able to point out that this procedure not only eliminates terminological disputes, it fragments research programmes by compartmentalising various aspects of theory. It remains a matter of dispute whether the new philosophy of science or just common sense is required to dispel the attractions of operationalism.[6]

In the case of the social sciences the excesses of behaviourism and other positivistic reductions of social phenomena can readily be linked to a misleading philosophy which is vanquished by post-empiricism. But here again symbolic interactionists (see Chapter 5), ethnomethodologists and others maintained assaults on the false philosophy without assistance from Lakatos and Popper. As we have seen ethnomethodology may have gone too far in its anti-scientific attitude but it is difficult to assess just how significant philosophy has been in atoning for its own mistakes.

The second positive claim is shrouded in a negative admission. Lakatos sees the risk of 'legislative interference' in the case of advanced sciences. The point here is a significant one, that philosophies may be conservative since they are derived from our experience of science up to the present. A radical scientific theory (such as relativity and quantum physics were for some) could be philosophically disadvantaged just because the current philosophy was worked out while earlier scientific ideas were dominant. Lakatos seems to imply that while this may be true for advanced sciences it is not a threat to non-advanced ones since they presumably follow in the wake of successful science. Accordingly, the philosophy of science might serve the non-advanced sciences whilst only reflecting the achievements of, say, physics and chemistry.

Something of this sort might reasonably be said to have happened in the social sciences. Just as particular mathematical techniques have been handed down, for example from electrical theory to economics, an experimental approach could be said to have been passed from the natural to the psychological sciences.[7] The fact that this has happened in one or two instances is not enough to establish a generalised hierarchy of advanced and non-advanced sciences. Many of the mathematical and biological models introduced into the social sciences could reasonably be held to have promoted only conceptual weakness and hasty generalisation. And it may also be said that there are distinctive characteristics to each of the sciences. Particularly noteworthy in this respect is the historical nature of social science which, as has been mentioned above, allies it more with geology and evolutionary theory than physics or the typical representatives of advanced science. Although it is common habit of thought to place the sciences in a hierarchy there appear to me to be no persuasive grounds for accepting that the philosophical hand-me-downs of physics are automatically appropriate for social science. In so far as the other sciences are bound to follow the path of physics an historical philosophy like Lakatos' may be of direct use but there seems no more reason to believe in this historical convergence than in any other teleologism, such as the idea that all nations follow the same route to modernity. For this reason I would accept as a general proposition Lakatos' assertion that his philosophy of science is unlikely to be of considerable help as a practical guide to scientists.

While Lakatos expresses doubts about the utility of his philosophy for practising scientists he does not make the grounds for those doubts particularly explicit. His allusion to 'instinctive scientific verdicts' would almost indicate that he feels that intuitive assessments, beyond formal specification, lie behind skilled scientific judgements. What then might these intuitive assessments amount to? A number of answers are possible. It may be that, as Kuhn has recently suggested, scientists operate with a number of scientific values which are used flexibly in making assessments of theories.[8] Scientists may value predictive success but also simplicity, scope and economy. There would be no strict hierarchy amongst these values although predictive and explanatory power would normally be foremost. Thus a skilled scientist might pass the 'instinctive verdict' that the simplicity of theory or programme A outweighed the scope advantage of B. The preference could not be logically deduced or proven. If such nuances are central to scientific evaluations then it is certainly the case that rules of method would be of scant use to practising scientists.

Supporters of interest theory would argue that there are even stronger reasons for the impracticability of rules of method. Interest theorists, it will be recalled, claim that even if opposing scientists purported to hold the same values they would interpret these values differently. In the case of the two statisticians, which was described in Chapter 3, Yule and Pearson might both have been in favour of simplicity but for Yule simplicity would, I suggest, have been understood as simplicity of calculation whereas Pearson would have treated simplicity as referring to a simple development of current theory. The backgrounds of scientists may dispose them to esteem different values and to interpret those values discrepantly. Neither Yule nor Pearson could be said to have the exclusively correct interpretation. The few empirical studies which have been directed at scientists' interpretations of method seem to endorse this view even though not associated with interest theory. In a study of biochemists, many of whom expressed some familiarity with Popper's theory of science, it was reported that scientists generally saw themselves as involved in critically testing their own beliefs whilst they viewed other scientists as frequently uncritical.[9] They claimed that their own experiments put their beliefs to test whereas their opponents' experiments were built on the assumption of correct belief. This state of affairs was described by scientists on *both* sides of the dispute. Scientists appeared to be interpreting critical testing differently when applied to their own beliefs than when applied to their adversaries'. Here again a rule of method, such as submit your beliefs to repeated critical tests, is interpreted differently depending on the actor's position.

The final complication in the usefulness of Lakatos' model arises in the comparative evaluation of competing research programmes. Scientists will commonly have to make a decision between, at least, two viewpoints neither of which is their invention or fully under their control. They cannot individually carry out all the tests or studies which they would like to see performed. Accordingly they typically have to make a decision when both competing theories have some features which the other lacks: some advantages in describing granites for example versus utility in understanding limestones. Lakatos' model only dictates that the programme with excess predictive and explanatory power is to be preferred but how can this possibly be assessed in any everyday scientific situation?

Following on from Lakatos' admission quoted above, I have suggested that the application of post-empiricist methodologies is not as straightforward as we might have hoped. Lakatos' statement suggested only that there might be contingent, circumstantial limitations on this applicability. The subsequent objections that have been reviewed argue that a theory of method may be *in principle* inapplicable because the interpretative use of rules in any situation is extremely dependent on scientists' viewpoints and commitments. The evidence reviewed suggests that even if competing scientists were intent on a disinterested invocation of Lakatos' rules the application of specific rules to contested issues would be as equivocal as the issues themselves.[10] This is not to say that the philosophy of science is of no use. It may, as Lakatos says, save scientists from the clutches of worse philosophies which, for example, say that scientific knowledge must confine itself to observable entities, but this is a rather ignominious task for philosophy! It may also provide us with a general view of the cognitive operations involved in the elaboration of theoretical structures, and philosophy can from time to time become involved in substantive debates by clarifying the meanings of terms and the presuppositions of theories. It may do this and more, but it cannot provide direct procedural rules which guarantee scientific success.

The place of rationality in social scientific explanations

Earlier on, when discussing rationality as part of the substantive theory associated with post-empiricist views of social science, it was found that there was an incipient paradox which arose in giving rational explanations. The difficulty was that there was no criterion for deciding between an irrational act and an act which we had simply failed to grasp the point of. Laing's work on the understanding of schizophrenic actors was cited as an instance of this paradox: by choosing to see the patients' actions as rational and imputing unanticipated wants and beliefs to the people, Laing was able to offer an interpretation of hitherto incomprehensible behaviour. Since this procedure is possible, it poses a special problem for the Lakatosian philosophy which depends on critical testing of alternative explanatory programmes. Leaving aside for the moment the practical and sociological objections raised against Lakatos' philosophy, it offers in principle to adjudicate between, for example, the thermal and chemical theories of rock formation because it is indifferent to which interpretation is correct. Both theories are appropriately scientific, both are naturalistic and cohere with presuppositions about the constancy of the natural world, and it does not affect the philosophical theory whether one is correct or the other. However, post-empiricism is not similarly indifferent to social scientific theories of action since the philosophical programme demands that people behave, and can have their actions explained, rationally. The philosophy, or meta-theory, proclaims that all theories - including theories of social action - should be subject to critical scrutiny while at the same time the meta-theory presupposes a certain theory of social action. The meta-theory, one might say, has a special interest in the correctness of certain tenets of social science.

This is clearly an involved issue and it is worth taking a little time to clarify it by indicating some of the things which I am not claiming. A particularly strong version of what I have said might be taken to mean that rationality is merely a name given to a restricted number of decision mechanisms or to a set pattern of values and that a preference for these is enshrined in the philosophical theory. The charge would then be that post-empiricism is viciously circular. An approximately analogous instance would be a theory of bird-nomenclature which defined birds as the set of flying creatures and avoided the issue of flightless bird-like creatures by the simple expedient of calling them 'flirds' and establishing a separate theory for them. In the course of inquiry one would anticipate that the category of birds would be examined more imaginatively than this. I am not suggesting that post-empiricism defines rationality in such a restrictive way as this since, according to Lakatos, scientific method is allowed to develop and adapt with the progression of knowledge. Nor am I claiming that post-empiricism necessarily has a conservative or static notion of rationality. The techniques of game theory, probability, the mathematical modelling of decision-making and economists' use of devices for demonstrating the aggregated effects of individual choices are all significant and interesting ways of elaborating the simple notion that it is rational to pursue one's wants in the light of beliefs to maximise one's expected satisfaction.[11] As was stressed in Chapter 2 these elaborations of the simple notion of rational action are not illegitimate rationalisations since all theories are inevitably developed through *ad hoc* adjustments to the fundamental core. The post-empiricist defence, that the theory of rational action is the only available research programme in the field and therefore must be supported even through *ad hoc* modifications, I acknowledge as correct in principle.

Turning, then, to what my argument is: the peculiarity of the rational theory of action is that in a significant sense there has never been a viable alternative. The apparent alternatives, like functionalism (see Chapter 2) and behaviourism, are not regarded as having been fought off in a Lakatosian struggle between competing empirical research programmes; rather they were undermined on abstract, 'in-principle' grounds. The unchangingness of this theory of rational action is perhaps indicated by the continuing traditions of moral and political philosophy based around the actions and choices open to the rational individual. This state of affairs is further acknowledged in the commonsensical complaint that whereas philosophy and politics students can still study, say, Aristotle and Machiavelli their natural scientific contemporaries are of no use to present-day students of the sciences.

The significance of this allegation can be appreciated through an examination of the concept of 'Ideals of Natural Order' introduced by Toulmin in his study entitled *Foresight and Understanding*.[12] According to Toulmin's analysis, the history of scientific ideas is characterised by changes in scientists' conceptions of what the natural behaviour for matter is. Thus for Aristotle it was self-evident that force was necessary to maintain any object in motion. It required, say, the constant exertion of a horse to keep a cart going; as soon as the traction was removed the motion began to slow. It was natural for bodies to come to or be at rest and, accordingly, being at rest or approaching it was self-explanatory. Motion of all sorts required explanation whereas in an important way being at rest did not. Similarly, Galileo's theory of matter and motion took certain cases as self-explanatory:[13]

> Think of a ship (say) on a calm sea, and imagine the resistances to motion progressively reduced, until we could neglect them entirely. If that were to happen, said Galileo, the ship would retain its original motion without change. If it had originally been at rest, it would remain at rest until some outside force started it moving; while, if it were originally moving, it would go on travelling along the same course at the same speed until it met an obstacle.

Here again some pattern of behaviour is seen as the normal course of affairs in need of no special explanation; only starting to move or coming to a halt needs to be accounted for. By Newton's time even this was seen as insufficient since far from viewing a curving path across the oceans as entirely natural, Newton argued that travel in perfectly straight lines was the self-evident course – a course which would take the ship at a tangent from the earth's surface without the intervention of gravity's external force.

This argument about the changing models of natural behaviour must in many respects appear familiar: it corresponds very closely with Kuhn's ideas on paradigms and the exemplary instances which they contain (indeed this is no surprise since Toulmin too introduced the term paradigm). But Toulmin is chiefly concerned to make one primary point, that in the development of science (whether that development is rational and cumulative or not) major changes have occurred in what we regard as the natural behaviour of physical objects. This suggests that an important area of innovation has been the imaginative capacity to re-interpret cases that were formerly taken to be self-explanatory.

A second aspect of Toulmin's claim is that at any particular point the self-evident case – corresponding to the ideal of natural order – is beyond explanation. There are no more fundamental operations by which it could possibly be explained. The issue is in fact neatly expressed by the term 'self-explanatory': there is something paradoxical in the idea of anything explaining itself. This proposal, that ideals of natural order are completely above investigation, is significantly exaggerated as is shown when Toulmin extends the notion beyond physics to everyday thought processes where, he suggests, something similar operates. A car driving smoothly down the road might be heard but we would generally have no questions about it. On the other hand, if its motion was erratic and accompanied with a great deal of noise we would probably wonder what was causing its behaviour. In this case, however, more could readily be said about the ideal instance of the smoothly travelling car – at least about the operation of the internal combustion engine on which its smooth functioning depends. Equally in the sciences one can ask about the sub-atomic processes which underlie the natural behaviour of molecules or photons of light. And some research projects are explicitly directed at trying to overturn the present limits of their discipline – theories of gravity or magnetism for example. That Toulmin's most extensive example comes from physics, which is often regarded as the most fundamental of the sciences, suggests that the strength of his conclusions may derive, in part at least, from the examples selected.

It does appear, though, that Toulmin's analysis is usefully applicable to the position of social science albeit in a perverse way. For one thing the model of natural behaviour does seem to be extremely unchanging. Second, and perhaps more striking, rational action is strongly self-evident and we have no means for further explaining it. But to identify a common element in social explanations is not at all to pretend that all sociological theories or hypotheses are the same. To take just one obvious example there are conflict and consensus interpretations or economistic and cultural interpretations of society and social order. However, in each of these cases the method for understanding the conflict or the consensus or whatever is vitally similar. Good reasons are provided why people would rationally want to behave in the way that the theories suggest. Consensus theories, roughly speaking, argue that people will accept inequalities and inconvenience for the sake of the advantages brought by communal life whereas conflict theorists maintain that the inequalities and exploitation are so intolerable as to motivate people to object. The theorists disagree, often very strongly, over the wants and desires that people have (their 'philosophical anthropology' as it is often termed) but neither side doubts that if only people's wants and beliefs could be identified their action would be intelligible. In a similar manner Galileo might have been unsure what the forces acting on actual ships were but, no matter what the marine disputes were, it was accepted that the ship's motion was generated out of the balance of prevailing forces.

This idea can be expressed also in terms of the concept of a regulatory principle that was introduced in Chapter 2. In evolutionary biology the morphology of organisms is inspected to find good, adaptive reasons why they have the form they do. Equally, Galilean mariners would have sought the forces which gave the ships the courses they had and their concepts of the force of waves, tides and winds would have matched this fundamental commitment. Social scientists use a regulatory principle of rationality to comprehend whatever behaviour they find people involved in. But whereas this mental procedure is indifferent to the constituents of the natural world so that successive regulatory principles can replace each other, the procedure itself depends upon rationality and is therefore quite ill-suited to detecting anything other than rationality behind human activity. The social sciences, I suggest, are characterised by an unchanging ideal of natural order. For this reason the Lakatosian methodology is at best only partially appropriate.

The great variety of social scientific explanations of many areas of social life, of industrial conflict and frustration at work for example, is attributable to the fact that by changing the constellation of people's supposed wants and beliefs an enormous variety of stories about social action is compatible with the regulatory principle of rationality. Amongst ethnomethodologists it is sometimes said that conventional sociological accounts or explanations are 'no news'. The sense in which social science is 'no news' is precisely in relation to the regulatory assumption of rationality since, by artful re-arrangement of their subjects' wants, sociologists can easily produce plausible explanations of any action. Once the trick is learnt, so to speak, the rest is (comparatively) easy. This point is sometimes used as a criticism of certain sociological perspectives but it is a facile criticism in that it is exhausted as soon as it has been accomplished once: thus D. Smith provides an ironical 'guide' to doing ideological analysis in which she points out that all that needs to be done is to establish some plausible rational connection between the interests of a social group and aspects of the knowledge to which it subscribes.[14] Yet it is unclear whether the comment is seen as invalidating all ideological analyses or whether it merely sums up the problems against which all analysts of ideology struggle. Less abstractly, Cicourel has shown how sociologists derive straightforward answers from subjects by interpreting open-ended remarks in the light of likely underlying motives.[15] No matter how ambiguous the replies, analysts can use their assumption of rationality to derive data which can then be used to formulate a conventional explanation of action. The criticism is misplaced because it implicitly suggests that analysts should be doing something other than this but if the ideal of a rational natural order is unavoidable then the implicit recommendation is futile.

This issue also feeds directly into the interminable complaints about the triviality of sociology. On occasions these are probably justified in a straightforward manner. If someone is researching the connection between unemployment and depression and finds the two correlated this is not going to surprise anyone (although the degree of connectedness might be surprising). But even when the research is not 'obvious' in this way, by virtue of the regulatory principle of rationality, sociological claims are fated to make sense. Since explanations of action ultimately invoke reasonable grounds as the explanation they cannot but appear obvious, or something very much like it, in retrospect.

Everyday and scientific knowledge

We have seen how under the influence of Lakatos and Popper post-empiricism is concerned with rationality as it appears in scientific method and, particularly, with the choice between scientific theories. However nothing that post-empiricists have said indicates that scientific knowledge alone is described by their theories. On the contrary Lakatos wants the history of science to be a rational undertaking as well and, as was mentioned in Chapter 2, the structure of a research programme is also found in alien cultures and even in witchcraft practices. It appears that what Lakatos has outlined is a general depiction of rationality rather than anything peculiarly scientific and this is much as we would expect since there can hardly be a cognitive gulf between science and prescientific ways of thinking about the natural world.[16] And if the attempt is made to restore the special qualities to science by saying that it alone is a methodic pursuit of answers then this too must be false in the light of the dogged reasoning of detectives, tax officials and the like. As long as we think of science as concerning the natural world this issue poses no special problems since most lay people are not concerned with, say, biochemistry or geophysics. But in social science there is a definite overlap of interests since for example both criminologists and detectives presumably share a passionate interest in the origins of criminal behaviour. Each profession will be equally likely to have its favoured explanations and it become urgent to ask whether social scientific theories are preferable.

At first sight post-emipiricism is ideally suited to answering this question because it is centrally preoccupied with the comparative evaluation of competing theories. But as we have seen in the first two sections of this chapter post-empiricism is actually weak in directing us in any practical situation of theory evaluation. Its recommendations are if anything more liberal than our intuitions about which theory we should prefer. There is, not to put too fine a point upon it, no proof that social scientific theories are better than everyday ways of thought and moreover no way open for Lakatosians to prove it.

It is not that Popper and Lakatos have not tackled this issue; indeed there is a fair degree of material on it which will repay some attention. As posed by post-empiricists this issue is described as the problem of the demarcation between science and non-science. Popper's original formulation of the theory of falsifications was directed as much at the demarcation question as at a theory of scientific practice.[17] His solution was that disciplines were scientific if their supporters could and did subject the tenets of the discipline to critical test. With the realisation that bona fide scientists did no such thing and with Lakatos' argument that they need not even try to do it, this solution lost a lot of its immediate value although it did not lose its appeal for post-empiricism. When Lakatos addresses the same question his answer is along similar lines: that a discipline is scientific if it does as much as is reasonable to subject itself to critical test. What has survived intact from Popper's handling of the issue is the sense that the two cardinal offenders against which the demarcation should be effective are Marxism and Freudianism. Lakatos, like Popper, can accept that certain disciplines might not want to be scientific and that they can establish their own rules (say, canons of accomplishment in landscaping or art criticism) but Marxism and Freudianism pose as scientific while being pseudo-scientific. For Popper this was relatively easy to establish since, for example, famed Marxist predictions had not been realised: there had been no pauperisation and the middle classes had swollen rather than disappeared in a dichotomisation of class. Accordingly, Popper felt that Marxists should trade in their social theory for a better one. Instead they invoked *ad hoc* reasons for the postponement of the proletarian uprising and explained the lack of political activity as due to false consciousness. They mimicked the form but not the procedure of science.

Lakatos cannot pass this judgement since he acknowledges that the avoidance of falsification by *ad hoc* modifications is quite legitimate. Yet he equally wishes to classify Marxism and Freudianism as pseudo-sciences. His grounds for this are not, however, made explicit, although two sorts of argument are put forward: that Marxism is crookedly unfalsifiable and that it is a rotten theory. Thus:[18]

> It is always interesting to ask a Marxist, what conceivable event would make him abandon his Marxism. If he is committed to Marxism, he is bound to find it immoral to specify a state of affairs which can falsify it.

> Has, for instance, Marxism ever predicted a stunning novel fact successfully? Never! It has some famous unsuccessful predictions.

In the light of sociological studies of scientists' behaviour in disputes one might merely conclude from the first argument that Marxists are more honest than most other people in acknowledging that they are not angelically tolerant. The argument certainly fails to establish its aim. But it is a bad argument for at least two further reasons: some Marxists do change their minds and, secondly, scientists can often change their detailed beliefs whilst adhering to the same general school. The second argument could be equally provocatively posed of (scientific) Darwinism, at least up to the time of biochemical genetics. In any case just because Marxism happens (perhaps) to be a bad theory that cannot make it unscientific for that reason alone. Lakatos seems to mistake the unscientific behaviour of some Marxists for a proof that Marxism is unscientific – as though say a few unscrupulous geologists could discredit their subject's scientific standing.

Rather than ascribe the paucity of these arguments to mere political ratiocination I suggest that there is a deep peculiarity in the structure of Marxism and Freudianism which accounts for, but does not justify, Lakatos' antagonism. As was mentioned in Chapter 2 the regulatory principle of Marxist studies is to inspect social institutions for the ways in which they subserve the interests of the dominant class. Supplementary claims about the possibility of false consciousness and ideology accommodate and account for apparently perverse political developments. Ideology and false consciousness operate also to explain why actions may often not be explained in terms of people's wants and beliefs. A precisely similar procedure underlies Freudian explanations: actions, especially compulsive ones, are scrutinised in the light of a regulatory principle of attempts by the id to evade the censure of the superego. Moreover, actors' own wants and beliefs, as consciously perceived, are once again inappropriate as explanatory features. Self-deception corresponds to ideology and false consciousness in the Marxist schema. The source of Lakatos' unease with both Marxism and Freudianism I suggest is twofold: on the one hand it may appear that in both instances the regulatory principle is used uncritically. Marxist analyses can all too readily depict every development of society as conspiratorially favourable to the dominant classes while Freudian accounts can easily present every action as imbued with deeper and prior motivations. In the unrelenting use of these regulatory principles lies the justification, if any, for the view that these two disciplines are unscientific because they evade critical test.

The second source of unease is rather more significant. Both Marxism and Freudianism demand that the analyst is methodologically dependent on rationality to disclose the 'truths' concealed by ideology and self-deception. The analyst must be able to see that racism is economically beneficial to capital or that the image of the horse expresses Little Hans' deepest anxieties. Thus like the model of the actor in post-empiricism and in Lakatos' theory of science, rationality is of fundamental explanatory importance. But the subject of that rationality is vitally different: for Marx it is the history of class antagonism and for Freud it is the constellations of the psyche. (For functionalists, it is worth noting, it would be society itself.) Thus the gravest opposition offered by Marx and Freud to Lakatos is not their methodology but their substantive location of rationality. Indeed, Lakatos is just as unrelenting in the application of his regulatory principle of individual rationality to social action as Marx and Freud are in the use of theirs.

This suggestion about the identity of the element in these theories which Lakatos finds unacceptable is indirectly confirmed in the case of evolutionary biology. As was mentioned in the discussion in Chapter 2, the regulatory principle of biology is to treat every feature of the organism as the product of selection. Thus the analyst seeks in every case to find a rational connection between the organism's features and its environment. The feature is considered unexplained until some such connection is detected. Much to the alarm of many scientists, Popper originally held Darwin's theory to be unscientific. The fact that Darwinism, like Marxism and Freudianism, depends fundamentally on the relentless application of a regulatory principle and that this principle demands that the analyst make rational connections strongly suggests that these are the ingredients of the syndrome Lakatos and Popper call pseudo-science.

This discussion of Popper and Lakatos has shifted away from the question which introduced it, the question about demarcation, and the connecting stages can now be sketched in retrospectively. The search for a demarcation criterion must be pronounced unsuccessful. The criteria supplied are not expressly aimed at separating everyday from social scientific knowledge nor can they be readily adapted for this purpose. What they do seek to exclude are the 'pseudo-sciences' of Marxism and Freudianism but the attempted condemnation of these research programmes is unsuccessful. What does emerge, however, is that the alleged pseudo-sciences share the property that they depend methodologically on identifying rational action, but rational action which is not performed by the customary human agents. Instead the postulate for methodological purposes is of rationally acting classes maintaining dominance, ids evading censure and species fighting for survival.[19] These disciplines also share the property that they pursue their regulatory principle relentlessly – no institution is deemed free of a class basis, no act is free of subconscious meaning and no organ is free of survival value – but ironically this is true of the post-empiricist theory of social action too: no culture, as has already been said, would be regarded as irrational. If it is this indissoluble attachment to a regulatory principle which makes disciplines pseudo-scientific then this is true of post-empiricist sociology and of Lakatosian history of science also. In a sense, because of the unchanging ideal of natural order in social knowledge the label pseudo-scientific for all knowledge of the social is virtually justified. The label, however, tacitly implies that this position could be improved – I consider it cannot. In this narrow sense sociology cannot become any more scientific.

It should also be noted that it is not being proposed that Marxism or Freudianism is as correct as evolutionary theory. In so far as Marxism implies a real commitment to the idea that institutions come into being for the sake of the ruling class it is as flawed as functionalism. But in so far as Marxism consists in the tendency to inspect social institutions primarily for the consequences they hold for varying classes then, for the industrialised world at least, it appears as reasonable a regulatory principle as the idea that actions are the outcome of subconscious struggles or the freely adopted actions of wilful individuals.

Reasons for action and accounts of action

As described above, ethnomethodological commentators sometimes adopt the view that the theories and interpretations propounded by sociology are 'no news'. More formally expressed, they hold an attitude which has come to be known as 'ethnomethodological indifference', claiming that the stories of sociologists are no different in quality to the narratives spun by ordinary members of society. All such stories are held to be based on indexical methods, like the 'documentary' method, and therefore of no general validity. To this supposedly inadequate sociology they offer an alternative: the study of member's reasoning procedures. On this interpretation all explanations of action are regarded as accounts – as described in Chapter 1 – that is, as ways of making action intelligible in the current context. According to Garfinkel we cannot discriminate actions from accounts of those actions. For him, in fact, accounting *is* doing.[20] We cannot get to the real explanation of, or reasons for, an action and must give up the attempt. All that can be done is to study, as it were, the anatomy of accounts. In this way ethnomethodology appears to offer an alternative to the post-empiricist view and also to solve the problem of separating social science from everyday knowledge.

On a strict Garfinkelian view ethnomethodology is quite unlike everyday people's concerns.

However, as a number of critics have been eager to point out, this is not the end of the story.[21] For one thing, even if ethnomethodologists were correct, this argument does not remove any of the immediacy of the everyday questions which sociology originally set out to answer in an improved way. Ethnomethodology does not remove the urgency of doubts about the sources of poverty or inequality. Moreover, if motives and reasons which purport to explain action do not in fact do so then surely the obligation is to find out what does rather than to study how these worthless reasons are composed. In significant respects this criticism is exaggerated, ethnomethodology does not claim that reasons cannot explain actions because something else does, rather the argument is that particular reasons cannot be shown to lead to actions since identifications of reasons are always based on revocable imputations. But even so, the critics' assertions can rightly be seen as a challenge to overcome the obstacles to knowing how actions come about - we are all too sceptical of a priori proofs to take ethnomethodologists' word for this impossibility. Furthermore, the suggestion that accounts should be understood as depictions of action appropriate to the prevailing circumstances tacitly suggests a reason for action since accounts are actions performed in speech, speech acts. There is at least the possibility that accounts are ways of making past actions appear appropriate to the present moment, a speech act which is only plausible if we suppose that people are generally motivated to save face or something similar.

The final, and strongest, objection to the attitude of indifference is a methodological objection. Accounts are not treated as isolated occurrences, rather various types of account are recognised, including greetings, compliments, and apologies. Yet to recognise that two accounts are of the same class presupposes that the account indicated the speech act of which it forms a part. A greeting or compliment is not composed of mere words but of actions undertaken through words. Thus any systematic work on accounts demands a classification scheme which depends on recognising, at least in general terms, patterns of action. Of course accounts are not determinate: not all complimentary phrases are compliments but at least some compliments must really be so for the catgory to have any utility at all. This series of issues requires further treatment, which it will receive along with empirical illustration in the next chapter.

The immediate point to be drawn from this brief discussion is that the ethnomethodologists' claim that we cannot have knowledge about reasons and motives is crucially ambiguous. It is not, as is sometimes suggested, that we should simply forgo such knowledge - even less can such knowledge be substituted by ethnomethods. But since social knowledge appears to be 'pseudo-scientific' in the sense established above (that we can always find some rational reason for action and in many cases several viable reasons) then we have at least grounds for hesitancy about imputing motives and reasons. Naturalism is here to stay – indeed everyday knowledge seems itself to be naturalistic – but there are great procedural difficulties in being confident in the identification of particular explanations in terms of actors' reasons. Under these circumstances something akin to an anti-naturalistic study of how practical reasoning operates, related to the kind of analysis described as studies in scientific work, will assist in understanding how reasons and motives are imputed and understood. Sociologists will not wish to give up the task of providing narratives about social action, and these narratives will inevitably depend on an imputation of rationality and the ascription of wants and beliefs. But since diversity can only be anticipated of such work, studies in how imputations and ascriptions are made must come to play an increasingly significant role, not only in substantive studies in the area of microsociology, but in studies of methodology.

Conclusion

At the beginning of this chapter several impediments were recorded to the utility of the philosophy of science in guiding scientific behaviour. Its benefits for social scientific practice could also reasonably be doubted. Yet, it was argued that Lakatos' post-empiricist conception of science gave the best general depiction of how science might operate as a cognitive system (even though this system was unlikely to be unique to science) and that it gave a lead in breaking down the supposed bases for discriminating natural from social scientific practice. But since the theory of science was at the same time a theory of scientific conduct it provided a new basis for the distinctness of social science: social science is wedded to rationality in a way that is not the case for natural science since social scientists must impute rationality to their objects of study – even when bizarrely they study whole societies or classes in this way. This bond to rationality means that a dimension of change open to natural scientific theories is denied to social science. In some ways the latter is conservative. The inescapability of rationality means that in practice it is hard to discriminate between any number of viable explanations on each particular empirical occasion. This does not mean, as some ethnomethodological writing argues, that our concern with rational action must be superseded. With what could it be replaced? None the less, some of the forms of analysis to which this ethnomethodological objection has led will prove useful in understanding how our attribution of rationality functions. Inquiries which generate 'stories' or accounts about the social world must be combined with studies of reasoning. It is to examples of these two strategies of work that we turn in Chapter 5.

CHAPTER 5

Science and social research practices

Introduction

In this chapter the analysis of the philosophical position of the social sciences developed so far will be invoked to assess a number of predominant research strategies in contemporary sociology. The examples selected have not been chosen because they necessarily mirror the most famous theoretical traditions within social thought, nor, particularly, on account of their contentiousness. Instead they have been chosen to represent what I take to be among the most striking empirical procedures. Thus, for example, although conversation analysis is undertaken by relatively few people and has few philosophical advocates, it is undoubtedly a lively empirical tradition and is discussed here for that reason. The following discussion may strike the reader as atypical in another sense: philosophical commentators are frequently cautious about assessing empirical work because it is felt that philosophical considerations can only license general investigative strategies, and definitely not particular questions about the evaluation of data. Whilst I fully accept that philosophical analyses cannot routinely resolve empirical questions, my aim will be to take the assessment of sociological strategies as far into the research process as possible. It is precisely because post-empiricism has made the philosophy of science into a theory of scientific conduct that philosophic analyses ought to address the question of how research is to be performed.

Expressed in very general terms, my argument so far has been that the model of rational action, based on wants, beliefs and decision procedures imputed to the actor, is central to sociological analyses. This naturalistic scenario is however beset with pragmatic difficulties in the correct attribution of wants and beliefs and is suffused with a kind of cognitive inertia permitting no change of the ideal of natural order. It appears necessary to offer narratives or accounts about action but enormously difficult to certify any particular account. And as if problems of principle were not sufficient, social studies of science remind us that any cognitive closure that appears to be achieved may well have been driven by extra-scientific forces. In the light of these considerable obstacles to successful sociological work it will be necessary in each of the following cases to bear four questions in mind. It will first of all have to be discovered whether the approach is in principle science-like, and then whether it can practically generate science-like conclusions. Any non-naturalistic attributes will also have to be examined, as well as the 'philosophical anthropology' or, more specifically, the image of wants and beliefs and the reasoning mechanism attributed to the agent. The styles of research examined, in order, are symbolic interactionism, 'macro-sociology', statistical analysis, ethnomethodology and conversation analysis, and the sociology of knowledge.

The explication of interaction

As mentioned in earlier chapters, symbolic interactionism and related work in social phenomenology has featured prominently as one of the currents of social research which claim to offer an alternative to H-D naturalism. This antipathy to naturalism appears to derive from two sorts of sources. On the one hand, it arises because symbolic interactionism emphasises the symbolic and meaningful facets of social life in preference to physical and often quantifiable social variables. Additionally, research interest has frequently been directed to the point of view of minority or 'underdog' groups. Thus both in its subject matter and in its epistemology it seems to express an anti-establishment outlook. However, these two attributes have no necessary connection: even the ruling class or the humdrum bourgeoisie can be studied from an interactionist perspective, as Goffman famously demonstrated.[1] In order to assess the interactionist research strategy more systematically it will be necessary to keep these threads distinct, and this can best be achieved through an examination of specific interactionist studies.

A well-known and useful study with which to begin is the analysis of marihuana use by Becker.[2] He notes that the majority of previous studies had asked 'why do they do it?'; but it is his contention not only that the kind of answers (in terms of psychological weaknesses and abnormalities) are wrong but crucially that the question is wrongly posed. Doubtless many present-day readers will share Becker's intuition that the question is bizarre: it is now felt that marihuana use is quite probably as reasonable as smoking and not in need of some special explanation. This rather 'quaint' aspect of the study conceals an important issue to which I shall return shortly. From his point of view Becker was concerned to oppose psychological or physiological accounts of marihuana use (clearly science-like accounts) and to replace them with symbolic accounts. In this way his project appears definitely anti-naturalistic; but before this assessment is passed the actual details of his analysis must be considered.

Becker breaks down the process of becoming a marihuana user into a number of stages, examining in turn how the novice user learns the preliminary techniques (how much to smoke, how deeply to inhale and so on); how the user comes to perceive the effects of smoking and to attribute them to the drug; and how these effects are understood as pleasurable. Taking this last aspect for example, Becker is insistent that the consequences of taking marihuana are not in themselves pleasurable. Indeed he cites testimonies suggesting that the experience can readily be perceived as deeply disturbing:[3]

> It started taking effect, and I didn't know what was happening, you know, what it was, and I was very sick. I walked around the room, walking around the room trying to get it off, you know; it just scared me at first, you know. I wasn't used to that kind of feeling.

However, novice users can quickly come to focus on particular aspects of these alien experiences and to cultivate those features which are seen as pleasant. Becker argues that advice and encouragement from experienced takers is usually necessary to attain this degree of discrimination. Only encouragement from members of a culture which values experience of, say, dislocation and which expresses this as a prized form of liberation can make the brute physical effects pleasant. The use of marihuana depends on a symbolic environment in which users can be socialised into ways of perceiving. And in a similar fashion, only in such a symbolic culture could successful drug use be learnt and the drug-induced experiences conceived of in a regular way since the culture supplies the vocabulary (of 'getting high' or appropriately 'happy') necessary for the basic accomplishment of recognising the correct effects of the drug.

At an analytic level, Becker concludes that the symbolic dependence of marihuana use demonstrattes that the customary question, which seeks a cause for 'drug abuse', is false. This is because, in the course of learning the symbolic culture surrounding marihuana, the individual[4]

> develops a disposition or motivation to use marihuana which was not and could not have been present when he began use, for it involves and depends on conceptions of the drug which could only grow out of the kind of actual experience detailed above.

Thus Becker invokes an epistemological argument: a propensity to enjoy the drug or the kind of experiences associated with it could not have been the cause of the deviance since these attributes were not present in the individual intially. The symbolic explanation appears to be of a kind which differs radically from any scientific conception of cause. Yet this argument is crucially ambiguous: at one level it appears to be a methodological objection to the original formulation of the anxiety about marihuana *per se*; whilst at another it seems to suggest that the answers to which the original question leads are simply false and that a different kind of mechanism actually leads to drug use. On this latter interpretation, his argument implies that anyone who enters a particular subculture can readily come to regard marihuana as acceptable and enjoyable. Thus he replaces a physiological mechanism with a sociological one and, moreover, side-steps the question of why specific individuals enter drug-using cultures in the first place. His overall claim could, I suggest, reasonably be paraphrased as follows. For environmental or even quite chance reasons certain individuals are exposed to drug-using subcultures, whereupon through the acceptance of the new symbolic surroundings they can readily learn to be the 'kind of people' who enjoy taking drugs. Now, the accidental nature of their introduction into this state means, in a certain limited sense, that it is useless to inquire why these particular individuals take marihuana. But the hopelessness of the question is not a matter of principle or epistemology, but a matter of circumstance. And this is further reflected in the present-day popularity of marihuana: it is not *epistemologically* nonsense to ask why certain people take the drug any more than it would be to ask why some people take milk in coffee, rather the widespread nature of the practice makes it unlikely that any single precipitating cause could reasonably be detected. Hence a mere change in the popularity of the social practice makes the original (quasi-biological) formulation of the question appear ridiculous even without the help of Becker's symbolic analysis. Becker's apparent anti-naturalism comes from the fallacious argument that if a science-like question cannot reasonably be answered, it is because no scientific inquiry should be made. But, as was shown in Chapter 1, the complexity of phenomena may make science impracticable without making it epistemologically improper.

It is not, however, always the case that symbolic interactionism denies the significance of the more usual or 'establishment' questions about social action. In his recent work, *Frame Analysis*, Goffman offers the admission that he makes:[5]

> no claim whatsoever to be talking about the core matters of sociology – social organization and social structure.... I personally hold society to be first in every way and any individual's current involvements to be second; this report deals only with matters that are second.

Whilst perhaps reserving some irony in his tone – for these matters are commonly not second to the people involved – this acknowledgment at least offers to take seriously a form of criticism often levelled at interactionist work: that, for example, the details of the supply and merchandising of marihuana, which presumably make someone wealthy, are overlooked. Goffman makes clear that his inattention to such things is not a matter of methodological or epistemological necessity but a mere reflection of his personal choice.

At one level, this choice takes him into what he regards as 'neglected situations', like face-to-face interchanges, which have received too little analytic attention but it also has a more theoretically significant pay-off. Throughout his work he argues that people are not only instrumentally concerned to achieve whatever action it is that they are undertaking, they also take pains over the actual doing of the action. People are commonly interested in both what they are doing and how adequately or well they are seen to be doing it. In this sense, therefore, secondary matters attain a certain prominence. Indeed for many individuals these two sets of considerations can fuse for, as he reports in *Stigma*, people with spoilt identities (that is, people whose stigmatisations obtrude on regular social interaction) are frequently concerned with accomplishing things unremarkably as much as with accomplishing them *per se*.[6] And under circumstances of privation, as recorded in *Asylums*, with very few tasks and opportunities in life, people may promote the details of interaction to the principal goal of their existence.[7] Such matters may come second but it is often a very close second.

While his work can readily be viewed as a collection of entertaining and enlightening stores about people's interactive strategies and their efforts at self-presentation, it is important for present purposes to ask what these findings amount to at an analytic level. In the course of his work, Goffman makes three main sorts of analytic claim. Firstly, he documents, particularly in *Relations in Public*, the extent to which people possess and make use of subconscious and inexplicit knowledge. Thus, he suggests that, for example, on entering a train compartment we can generally tell from the way that people are sitting and the way their possessions are deployed, whether the vacant seats next to passengers are available or not.[8] In the terms of *The Presentation of Self in Everyday Life*, we can read the signs that people 'give off' without being explicitly aware of what aspects of their behaviour we are responding to.[9] Secondly, as described above, he claims that people monitor and care about the appearances they generate: broadly speaking, that they take a great interest in the 'means' as well as the 'ends' of social action. Finally, and following closely from the second point, people interpret and evaluate action reflexively. By this I mean that actions seem not to be assessed in the light of some set of absolute standards but in the light of the actor's likely reasons for action. Returning to the train carriage example, someone entering a fairly empty compartment who then sits next to a fellow passenger may readily be regarded as attempting to establish a conversation. If, however, their leg is in plaster, it may be thought that they are sitting down to avoid having to move further from the door – even if they subsequently use the opportunity to develop a conversation. Actions are not judged simply by reference to rules or norms but by inferences about reasons and imputations of motives in the light of the current context. And of course this interpretative quality of action is open to manipulation: an injured leg can become the pretext for initiating conversations and, as Goffman records, thieves who mimic a bank's daily routine can blithely drive away as though they were bank personnel without raising suspicion.[10]

In terms of methodology, Goffman's studies are open to attack as unconcerned with representativeness. In one sense this is certainly the case – although he makes no attempt to conceal it – and many of the nuances and niceties of interaction which he reports are quite probably only mid-American traits. Accordingly, any claims about people's subconscious knowledge may be of very local validity. In spite of this, and aside from pragmatic arguments about the inevitability of concentrating on small numbers of cases, Goffman's work can be methodologically defended with reference to issues of philsophical anthropology and the essential role of rationality. This is not a theme which, it seems to me, Goffman himself develops but it is suggested by some of the claims in *Frame Analysis*. At their simplest, frames are 'schemata of interpretation' which underpin our everyday understanding.[11] As such the primary frames can be viewed as the basis upon which the reflexive imputations described above are made. He then proceeds to describe alternative, derivative frames which encompass activities like practising, play-acting and mimicking of the primary frames. For such activities, elements of the primary frame are suspended but a vital connection with other aspects of the primary frame is retained. Many of the jokes, diversions and obsessions of life are bound up with these derivative frames. With this analytic move, Goffman is able to suggest that whatever the content of a culture's primary frames, the structure of primary and derivative (or 'keyed') frames will persist. Frame analysis is a formal examination of the structuring of interpretation; it is not a theory about the content of any particular frames. Without fully accepting the notion of frames (which appear to diminish the importance of subconscious knowledge and the openness of people's interpretational skills), it can be argued that the main significance of Goffman's work is contained in his claims about how people interpret the social world and understand each other's actions, claims which are of a nearly philosophical nature – a status parallel to that of the post-empiricist model of the rational actor. Just as it has been argued that the model of rational action is beyond straightforward testing because it is presupposed in any interpretation of behaviour, so it also appears that the kind of reflexivity described by Goffman would be similarly presupposed. His work can be viewed as a contribution to the model of the actor and therefore not directly open to questions of representativeness.

According to the interpretation presented here interactionism is not anti-naturalistic when viewed in the light of the post-empiricist philosophy of social science. Becker's work may appear to be anti-scientific in that he turns his back on the question of who becomes a drug user, but it turns out that this amounts to a claim merely that the process is too random to submit to analysis. Goffman also evades such questions - perhaps out of a scepticism about ever getting meaningful answers - but pays lip-service to their importance. Moreover, such issues form a vital backdrop to the behaviour of spies, actors and industrial espionage of which he is so fond. Nothing in symbolic interactionists' formulations denies the possibility of a science of action: nothing contradicts the calculus of wants and beliefs which characterises the post-empiricist and Lakatosian model of action. On the contrary, Goffman adds new beliefs in the form of subconscious knowledge and the signs 'given off' by the body. More significantly, he introduces new wants, wants that are to do with the preservation of a self-concept and with the successful accomplishment of local details of interaction. With such a diverse and perplexing series of wants and beliefs it is not surprising that people should often be pre-occupied with efforts to present themselves favourably and with trying to work out what other people are 'really' up to. People's attempts to theorise each other's behaviour are thus fully compatible with a (modified) model of the rational actor. Game theory, it might be said, simplifies the complexities of action by assuming that people concentrate on specifiable ends. But if the doing of things and local interpretative requirements are significant in themselves, then action is far more complex than strictly rational, instrumental philosophical anthropologies suggest. Interactionism does not throw doubt on the naturalistic model in the way that ethnomethodology seeks to, it merely complicates the over-cognitive and over-instrumentalist model of the actor. In line with the curious status of the supposition of rationality in the hard core of social science, the contribution of interactionist work can be seen to be partially conceptual and partially empirical.

It should finally be noted that there is a tacit assumption that people's statements form a ready and unproblematic source of information. In Chapter 3 possible restrictions on the use of scientists' accounts as evidence were raised and this is in stark contrast to Becker's dependence on testimonies about marihuana-induced experiences. This question of the analytic significance of language will be reconsidered in a later section of this chapter. At this point it must remain a possibility that our language as much as our bodies and dress contains significant framing information. None the less, at a formal level, it appears from this analysis that Goffman's contribution to the extension and elaboration of wants and beliefs is comparable to the theoretic contribution of Freudianism. While Freudianism replaces the cognating individual for methodological purposes with a hierarchy of rationally self-seeking inner selves, Goffman substitutes the cognating individual with self-conscious agents concerned with interactional propriety and success.

Studies of social structure

In this section the status of macrosociology, understood in the broadest sense, will be examined. The conspicuous feature of interactionism which makes it appear initially incompatible with naturalism – its lack of representativeness – is absent in macrosociology which aims to deal with large numbers of cases and explicitly makes use of comparisons. Perhaps the major obstacle to seeing macrosociology as naturalistic is the issue of the regulatory principle already discussed in Chapters 2 and 4. The regulatory principle of functionalism, which involves inspecting every social institution for its contribution to the continuance of society, is significantly at odds with the post-empiricist view of the rational actor. Accordingly, naturalism does not easily license a science of social institutions and organisations which has structural variables as the fundamental explanatory items. Just as Lakatos' view of science builds an institution out of the behaviours of individuals, so also must a post-empiricist macrosociology have as its regulatory principle the rational behaviour of particular individuals. This need not amount to an extreme methodological individualism: as has already been suggested individuals' wants and beliefs can themselves be considered as arising from previous actions and macroscopic trends may therefore be readily discernible. None the less the reproduction of these tendencies must be understood as the work of individuals.[12] Within these constraints, however, a large number of macrosociological theories are potentially viable. The crucial point is that post-empiricist naturalism, whilst individualistic, does not prohibit a macrosociology.

Having thus dispelled the obvious obstacle to the development of a scientific macrosociology there remains a practical problem in deriving individual correlates of macroscopic concepts. If, for example, one is concerned with classes or class division it is necessary to establish some way of detecting class in the case of particular actors, that is, a way of operationalising the concept. The problem here is rather complex and is not made any easier by an attractive misunderstanding: it might be thought that the problems of operationalisation arise because the macro-sociological concepts are theoretical while individual attributes are factual. But this is not the case for, as we have seen, individuals' actions and attributes are also understood through the imputation of theoretical entities, like reasons and rational decision mechanisms. The 'translation' of macroscopic concepts into individualistic terms is demanded not by the need to ground the concepts in factuality but by the regulatory principle of individualism.

The difficulty begins with the supposition that for every concept - such as the rate of inflation or degree of class consciousness - there ought to be some corresponding measure. Just as it is assumed that in every electrical wire there is some voltage or other and that every flask of water has some definite temperature, it might well be supposed that every Western society should have some degree of class antagonism or anomie. Moreover, because there sometimes exist several competing measures for social concepts, like the various figures for the money supply or different measures of educational attainment, then it is possible to conclude that macrosociology cannot be properly scientific. But just how serious is this difficulty in obtaining operationalisations of social scientific concepts?

The practical significance of this issue can best be explored through the use of an example. One case which has attracted a good deal of useful discussion is the notion of wealth - a concept which has obvious significance in attempts to make international comparisons of inequality and in efforts to monitor developments towards equality within societies.[13] Clearly wealth, like warmth, is a vague concept, but the apparent success in constructing devices which operationalise temperature in a convenient manner poses a challenge to a naturalistic macro-sociology to supply measures of wealth. Although a great deal of work has gone into the development of such measures there remains a great variety of them. Among the crucially disputed issues are: whether wealth is to include the value of belongings (housing, clothes and so on), perks, company cars and other allowances and, if so, how they are to be valued; and a similar series of questions for stocks, bonds and other investments. The estimation of the spread of wealth in any society will consequently depend in vital ways on how the selection of these factors is made.

A further series of issues seems to be involved in the selection of these disputed measures. While some of the questions, about, say, the value of clothing, seem rather arbitrary others appear much less so. Thus, it is disputable whether or not pensions to be received in the future should count as wealth: at one level pensions manifestly bring increased spending power to those who receive them but future pension 'rights' are of no fixed value, depend arbitrarily on how long one lives, may be at the whim of future governments, and do not currently exist since pensions are commonly paid out of contemporary contributions. The inclusion of pensions tends to decrease the measure of the inequality of wealth and so may be favoured by 'conservative' analysts while their omission may appeal to radical analysts. Finally, other differences in the estimation of wealth have more clearly theoretical roots. Weberian analysts might be expected to regard the important aspects of monetary income, and hence 'real' wealth, as disposable income after necessary outgoings at specific stages of life. Writers from a Marxist perspective, on the other hand, would interpret only that wealth as important which contributes to the generation of new capital, while income in the form of pensions and allowances would generally be useless for this purpose. Indeed small sums of consumer wealth might be held to be necessary for stable capitalist production. However, even within these competing views there may well exist differences of detail in how wealth is estimated. There appear therefore to be different kinds of difficulty in operationalising macrosociological concepts: some practical and some theoretical. How then does the philosophical model developed above suggest that these should be handled?

Ironically post-empiricism offers the readiest solution to what, at first sight, might appear the severest of these difficulties. Conflicts such as that between Marxists and Weberians over wealth are easily accommodated by a Lakatosian model of competing research programmes. Indeed post-empiricism would lead us to expect that for advocates of either side wealth would appear to be a factual matter and that they would regard their opponents' views on wealth as a simple distortion of the true state of things. That such difficulties of operationalisation are not peculiar to social science can be illustrated by an analogy with the disputes in the late eighteenth century concerning the nature of light: was it made up of waves or particles? In this instance too, contrasting scientific views about the true nature of an everyday phenomenon existed. According to the post-empiricist interpretation, developments internal to these views of light, constrained by the hard core of physical theory, led to a swing towards the wave theory. In this way our understanding of light was shown 'really' to be a vague grasp of this wave theory. And, similarly, competing claims about what wealth 'really' is should be decided by internal elaborations of the rival conceptions of wealth and by their relative ability to account for, say, the way that social power derives from wealth and the manner in which wealth is concentrated over successive generations. Of course, this description does not indicate that the process will operate smoothly; for one thing it has been suggested that the wave/particle dispute was resolved in practice not by an intellectual success but through the internal politics of the French Academy of Sciences, where an institutional coup instated 'pro-wavers' in the foremost offices.[14] Such closures of debate could readily be envisaged in the case of macrosociology as well. Furthermore, for the reasons discussed in Chapter 4, it is not at all clear that one could settle disputes in a Lakatosian manner even if one sought to. But in general it is clear that the persistence of disputes of this sort does not render macrosociology unscientific or epistemologically questionable.

Turning then to the technical problems, by which I mean questions of how exactly Marxists or Weberians are going to measure wealth, it is informative to re-examine the comparison, sketched above, between the operationalisation of temperature and wealth. If it is the case that temperature has an undeniable technical realisation then so much the worse for wealth, and social science. But the technical problems are usually exaggerated by such contrasts. Any actual thermometer is going to be subject to manifold sources of variation: the mercury may not be particularly pure - indeed it could never be said to be *completely* pure - and thus not uniform in its expansion; there may be friction inside the thermometer, and if the object is at all large then one has to decide where to locate the thermometer. For these kinds of reasons it cannot be said that physical objects have some known temperature. Instead out attribution of a 'true' temperature is based on an idealisation. This idealisation may be more or less far-fetched but it must inevitably be an idealisation. On any specific occasion it can only be a matter of judgement whether the actual measures used are adequate or not; and as was reported by EPORists, the scientists studying gravity waves were unable to agree whether the available sets of apparatus could come near enough to the ideal version of a gravity wave detector.

On the face of it therefore it is not scientifically peculiar that macrosociology should experience difficulty in setting up measures, whether of inflation, anomie or whatever. But it is precisely at this point that the kind of argument (introduced in Chapter 1) associated with Winch and MacIntyre, about the role of meaningfulness, rises to significance. It will be recalled that (part of) Winch's claim is that only the prevailing culture allows actions to be properly identified. The macrosociologist, on the other hand, has to decide that certain individual acts count as instances of some larger tendency: that votes stand for opinions or that hours spent striking indicate the extent of work dissatisfaction. In the natural sciences where the objects of study do not participate in the same kind of interpretative work as the analysts, the scientist is able to offer an operational definition of some phenomenon and then, within the constraints of the surrounding physical theory, adjust the concept of the phenomenon *and* the operational definition. Although particular techniques for measurement only approximate to the phenomenon supposed to be being measured, there is no independent route to that phenomenon. In the long run therefore our conception of the phenomenon will come to be based on successful past operationalisations. In the social sciences, however, the ability of the subjects to engage in analytic procedures means that operationalisations can be challenged by the 'objects' of knowledge. This does not mean, as Winch seems to argue, that the local culture is uniquely correct but it does mean that the sort of 'dialectic' between conceptions of the phenomenon and operationalisations which is found in natural science cannot operate in social science.

This complex set of issues can be presented by reference to an observation made in Chapter 2, that realism and conventionalism are not as far apart as is generally thought since each research programme holds its hard core and central principles to be true for the moment. This can now be re-phrased as follows: we commonly have a realist conception of the idealised phenomena but these ideals are only realisable under conventionalised operational circumstances. The realist expresses the gap between our operationalisations of temperature and the idealised notion of true temperature, while the conventionalist demonstrates how the idealised notion is tempered by attempts to measure it. In the case of social science, the actors' world is partly constituted through their conceptions of that world. Accordingly, the realist trend in this 'dialectic' is stronger. It is not that epistemological rules prevent the sociologist from departing from the members' concepts but the element of judgement required is far more contentious when the analyst conflicts not only with other analysts but with subjects as well. So long as macrosociology is debated in generalised terms these issues do not arise because detailed judgements need not be made. But as soon as empirical work is undertaken in any depth the specific details become open to challenge.

One further consideration also exacerbates this problem. There is often a restriction on feasible operationalisation because of the artefactual or 'man-made' quality of the issues under discussion. For example, in the case of the sociology of science, citation counting – that is, a measure of the frequency with which a publication is referred to – is a widely available measure. It has been used to operationalise, among other things, the quality of scientific work and the reputation of scientists. In some general way it is quite feasible that good quality work will be most widely cited or that the work of reputable people will command a wide audience. However the sheer limitations of the available information mean that these distinct variables are monitored through their shadowy connection with the same index.[15] And given that scientists themselves are likely to be concerned with the readership of their articles, it is not surprising that these operationalisations are criticised by the subjects. This restriction on the practicability of macrosociology comes about not from a version of the Winchian argument; it arises because actors are also analysts – they too are involved in reasoning about each other's motives, reasons and beliefs in the reflexive way illustrated by Goffman.

To sum up, macrosociology has a naturalistic orientation and faces no epistemological disqualifications at the hands of post-empiricism provided that it does not resort to a crudely functionalist regulatory principle. Disputes between alternative research programmes will generate conflicting factual claims and the *ad hoc* development of theories. However, the major obstacle to macrosociology is not the persistence of disagreements between research programmes – these would at least encourage internal development – but the problem of establishing operationalisations in an object domain which is composed of analytic beings. It is in the forging of connections between the theories and the evidence that the greatest problems arise and although these appear to be only technical, they are imbued with a significance for the subjects which belies this. A remedy for the 'slipperiness' of this evidence is often sought in statistical analysis, and a brief discussion of statistics is therefore demanded.

Statistical presentations of social action

Within the social sciences statistics is commonly seen as a divisive issue. Two sorts of extreme opinion about the role of statistics have become established and the assessment of statistics encouraged by post-empiricism can conveniently be presented through a consideration of these extremes. The two, opposing views, generally expressed, are that (a) statistics is a simple extension of measurement and counting and that (b) statistical analyses are singularly inappropriate for the social sciences. To begin with the former, there is clearly something to the traditional argument that statistics, being a brand of mathematics, comprises only formal, or what is sometimes termed 'analytic', knowledge and is therefore uncontentious.[16] According to this interpretation, any statistical manipulation of mathematical information cannot generate new knowledge. It can only re-present what was formerly, and perhaps implicitly, contained in the earlier formulation. At its simplest, to know that two eggs plus two eggs makes four eggs is not to know anything about eggs. This elegantly simple view has its weaknesses – and I shall come to those shortly – but what it does express adequately is the idea that statistics are not radically different from more straightforward ways of counting. Therefore anyone who wishes to outlaw statistics from social science has to make a case for the acceptability of the (small) amount of counting involved in their own, alternative approach. If, for example, a conversation analyst wishes to talk about the length of pauses between speech turns, or if an EPORist makes a claim about the utility of measuring devices as closure mechanisms by citing a number of cases, then both analysts depend on counting as a valid and neutral means of summing evidence. It cannot be in the dependence on counting *per se* that statistics is inappropriate.

Where the 'analytic', formal satus of statistics can be questioned is in relation to the development of complex statistical measures and functions. As was recorded in the case of the measures of association pioneered by Yule and Pearson, two different statistics could be derived to represent 'the same' property. Such possible divergences indicate that the extensions of statistical knowledge are not simply made up of deductive frameworks with no alteration of meaning. Furthermore, the 'analytic' status of statistics should mean that any mathematic formulation would be equally tenable, yet it is frequently the case that statistical tests involve the assumption of some mathematically regular distribution of properties among the subjects tested – for example, a normal distribution of intelligence. And of course the formal possibility of talking about the cube root of one's class affiliation or the natural logarithm of one's wants is generally never adopted. Whilst the interpretation of statistics as merely 'analytic' does usefully demonstrate the extent to which purportedly 'non-statistical' approaches actually rely on counting and elementary mathematics, it is not correct to claim that statistics innocently represents the initial data.

At the other extreme, statistics is sometimes attacked as though it involved some form of epistemological commitment which would make it inappropriate for social knowledge. In so far as this idea is supported only by the suggestion that the kind of encoded answers or restrictive formats of questionnaires are not adequate for sociology then it is at best not proven. For although statistical inquiry may encourage elementary forms of investigative work – because coded answers are easily computed – it does not necessarily do so. Social classifications based on quite different sorts of criteria could still be subjected to statistical analysis. A second line of criticism, outlined in Chapter 1, concerns the freedom of human action. This argument, it will be recalled, was that there appears to be a conflict between an individual's ability to act as they wish and the statistician's proposal that some particular percentage of people act in some specific way. In terms of the view of human action developed above this objection can be met. Since action is understood through the rational assessment of (imputed) wants and beliefs, given any particular set of circumstances people with similar wants and beliefs will be presumed to respond similarly. The statistical element expresses the analyst's uncertainty about the particular distribution of wants and beliefs; given the presupposition of rational action there could be no uncertainty about what people would choose to do if their wants and beliefs were known. The statistical component arises not because action is intrinsically statistical or random but because wants and beliefs can be so various. Some natural processes are held to be inherently statistical (like radioactive decay) and in these cases the statistical description is the only conceivable one. In social scientific instances - or at least in the vast majority since it is just about conceivable that spontaneous phenomena occur in the social realm[17] - the presumption and methodological centrality of rational action means that statistical accounts are summaries of individual actions.

Having dealt with the abstract aspects of this issue and stated the post-empiricist position, the procedural aspects must be addressed. At some point or other the statistical results - whether for example of differential social mobility rates or of associations between class background and leisure pursuits - have to be interpreted and, as has been acknowledged with respect to work on social mobility, 'the intelligibility of the relationship will depend upon providing an "action story-line", even if of only an implicit or rudimentary kind'.[18] There are two practical aspects to this. Firstly, although post-empiricism assures one that statistical relationships are indicative of something, it cannot demonstrate that they do not result from artefacts. These artefacts may be just the influence of undetected features of the social situation in which the original data were gleaned - as ethnomethodological critics propose. Moreover, it may be that as Goffman argues people generally operate through the use of stereotypes; or - and this may amount to the same thing - that in the derivation of formal measures subjects are not able to call on the subconscious knowledge which they would employ in practical situations.[19] Investigations carried on through purely statistical means are also obliged to abstract information from the complex array of wants and beliefs which people have and are bound, in this sense, to be reductive.

With the notable exception of Cicourel, ethnomethodological critics of statistical studies argue as though the objections they raise were unanswerable. Yet the authors of questionnaires and similar protocols are careful to minimise these kinds of effect.[20] Some uncertainty will inevitably cling to these results but their major drawback is the extent to which the kinds of information susceptible of complex statistical manipulation cannot do justice to the vast arrays of wants and beliefs which may motivate the post-empiricist human actor. And this kind of simplification relates directly to the second drawback with the introduction of an 'action story-line'. Given the rational model of the actor with an indeterminate set of wants and beliefs, any statistical relation will be able to support more than one conceivable story-line. The problem, discussed in Chapter 4, that social science is bound to make sense adopts a particularly virulent form here, for with the general and high-level nature of many statistical relations a great variety of story-lines can be produced. Essentially statistical studies fall foul of the interpretative gap introduced in the discussion of macrosociology. The most crucial part of the analysis – the interpretation of the results, whether they be citation counts or mobility rates – is relatively unconstrained. Elaborate statistics cannot compensate for a very fluid model of the actor when this fluid actor is effectively at the 'hard core' of sociological theory.

To summarise, statistics is not incompatible with social knowledge. Indeed counting and measurement, in their elementary forms, are part of everyday social knowledge as much as of social science. But the kind of sociological work generally thought of as statistical, whilst offering the only way of approaching some large-scale phenomena and touching issues of great interest, is particularly prone to the generic weakness of a science with an unchanging ideal of natural order. The vital interpretative link between the statistical result and the action story-line lacks the decisiveness which the initial resort to mathematics had sought to instil.

Practical reason and conversational structure

The methodological arguments devised by Garfinkel and other ethnomethodologists have been introduced a number of times in the preceding discussion but this has chiefly been in the course of criticising other styles of work. It is now time to assess the ethnomethodological research programme. To a certain extent, however, the fact that until now only negative issues have been associated with ethnomethodology reflects the position in which these analysts find themselves. Beyond Garfinkel's initial 'experiments on trust', which sought to discover the working presuppositions of everyday thought, it is not at all easy to decide what the next state of inquiry can be. With such a strongly critical view of conventional sociological research it is difficult to discern a methodologically unobjectionable line of study. One attractive avenue is to extend further the analysis of the structure of everyday practical reasoning. On the assumption that people use, for example, the documentary method and employ *ad hoc* means for adjusting their conceptions it ought to possible to study these reasoning devices in greater depth.

Among the analysts to adopt this procedure, Pollner's work can be conveniently taken as a representative. In his studies of 'reality disjunctures' he examines in detail how actors cope with the apparent existence of conflicting accounts of reality.[21] For example, magistrates or jurors may be faced with competing versions of an incident on a motorway: a police officer insists that a driver was travelling above the maximum speed limit while the defendant maintains that they were carefully monitoring their speed and never exceeded the limit. In the light of these conflicting accounts the magistrate does not entertain the possibility that both stories can be correct, nor that the two parties inhabit different realms. Rather the preference is for an *ad hoc* adjustment of some circumstantial detail which redeems both accounts and the presumption of a common life-world. A possible solution is that the motorist's speedometer is faulty. People appear to have a preference for interpretations which preserve the orderliness and rationality of their surroundings.

Pollner suggests that sociological studies commonly overlook this cognitive agility and, when faced with a reality disjuncture, simply side with one actor or the other. In general terms, the analyst may decide that motorists are more likely to lie than are police officers or that the structure of promotion within the police forces encourages officers to make charges on marginal evidence. Neither resolution appeals to Pollner. Instead he argues that the object of analysis should be the formal strategies through which actors resolve disjunctures. One case which he analyses in particular detail concerns claims by employees at a textile mill that they are suffering from bites and stings caused by insects imported with a shipment of cloth. Inspectors called in to investigate these claims could find no insects. The resulting conflict about the reality of the insects was complicated by the protagonists' reasoning abilities for both parties could 'see' why the other party disagreed with them. For the inspectors, the insects were invented by the workers as an excuse for laziness while, for the employees, the inspectors denied the insects' existence in order to oblige them to return to work. In the long run however such disputes are resolved, but this is not because the insects are found or shown not to exist. Rather, 'the resolution of full-fledged disjunctures implicates what for lack of a better word we choose to call a politics'.[22] With this proposal that resolutions are attained not by the acceptance of evidence or arguments but by some form of social/political closure mechanism Pollner has endorsed an EPORist position. Once Pollner has moved this far he has surrendered the neutrality which he initially adopted and has begun to offer conventionally sociological accounts of *why* people reason in certain ways. He would have stayed closer to his original methodological intentions if he had followed the strategy favoured by analysts of scientific discourse who concern themselves with regularities in people's accounts.

It is equally noteworthy that Pollner's study of practical reason takes him no further than would have been predicted by a Lakatosian model of the social actor. The kind of *ad hoc* adjustments he documents mirror precisely the methods that Lakatos recommends for the development of scientific theories.[23] At one level this could be interpreted as an encouraging sign for, apparently, both philosophical and empirical analyses of reasoning lead to the same image of the actor. More realistically, it can be viewed as indicating to how great an extent sociology occupies the middle ground between philosophy and empirical inquiry. That rational and empirical inquiries conform in their view of actors' reasoning displays both how unchanging this ideal of natural order is and how widespread this ideal is as an interpretative device in everyday reasoning. A simple documentation of people's reasoning strategies appears unlikely to generate distinctively ethnomethodological theories of human action.

The principal empirical outgrowth of ethnomethodology, conversation analysis, has pioneed an alternative approach to the potential conflict between the analyst's and the actor's interpretations – the conflict which Pollner sought to comprehend but which he could not finally avoid. By studying conversational interaction it is claimed that analysts can check their interpretations against the interpretative work of the co-conversant. Conversation analysis can usefully be traced from what Sacks described as the problem of description in sociology.[24] Accepting the range of ethnomethodological objections to conventional sociological descriptions of every kind, Sacks looked instead to people's own methods of description. Rather than describe any particular person as suffering from, say, alienation or 'having no-one to turn to', Sacks proposed that a procedural surrogate would be to examine when people can 'justifiably' say that they have no-one to turn to and so on. And in this context, justifiability is seen not as an epistemological question but as a question of what is accepted in the appropriate conversational context. Sacks sought to formalise the factors that led someone to be able appropriately to say, on telephoning a suicide prevention centre, that they had no-one to turn to. Starting from this fairly esoteric question the study of conversation spread in an attempt to formalise, over a far wider area of conversational issues, what the presuppositions and ground-rules of conversational interaction are.

The best-known studies in conversation analysis (CA) are not directly related to the issue of descriptions but to structural features of conversation.[25] Thus the most celebrated findings concern the organisation of turn-taking in conversation. Although conversations appear to arise from the spontaneous utterances of the participants, they are surprisingly orderly. People tend to speak one at a time and to co-ordinate the beginnings of their speech-turns very closely with the ending of the previous turn. Furthermore, some sorts of turns are naturally complemented by other turns: greetings match greetings, requests match acceptances/rejections and so on. These sets are termed adjacency pairs. CA is directed towards the elucidation of these ground-rules which are responsible for the local organisation of conversational action.

There are two striking features to these rules. Firstly they are normative and *interpretable* rules but not binding rules nor, like the rules of grammar, constitutive. Thus a greeting invites the recipient to return with another greeting and the absence of a greeting would not just be a neutral silence, it would be a silence interpretable as, say, a deliberate non-reply - or, more simply, a snub. Similarly, questions which are not countered with answers can be understood to have touched on a sensitive, complex or secret issue. CA does not maintain that the first part of an adjacency pair calls forth or determines the second part. It does, however, create a local context in which the second person's next act or utterance will be interpretable as situationally related to the question. It is for this reason that, when asked an embarrassing question, it is easier to pretend not to have heard than to be seen to ignore the question. Having once heard the question, one's actions will be open to interpretation in the light of the question's having been asked.

The second feature, already implicitly mentioned above, is the local nature of these conversational constraints In their turn, every speaker is concerned to act in ways which are, at least, sensitive to the interactive context. The efficacy of these local constraints can perhaps be illustrated with an analogy to parliamentary questions: questions to a minister cannot determine what ministers do nor prevent them acting in line with previous decisions, but they can generate a situation in which the ministers must either explain their actions or, if the question is evaded, be thought to have no good explanation. In the same way, conversation analysts are able to argue that the micro-contexts generated in interaction can so influence people's immediate wants and beliefs and thereby the course of their actions that no complex interaction or activity can reasonably be said to be completely predetermined or known with certainty in advance. In this way CA can be seen to contribute to the model of action in similar ways to Goffman and symbolic interactionism but through more formal means. Where Goffman had subconscious knowledge, CA proposes rules of conversation and where Goffman had reflexivity and social selves, CA has a succession of microscopic interpretative contexts.

Whilst theese elements can be construed as contributing to, and expanding, the notion of the rational actor and therefore to be compatible with a naturalistic social science, there is much at least in the attitude of CA that appears to set it apart from conventional notions of sociology. Advocates of CA are commonly resistant to claims to know anything special about the social order apart from what can be observed in transcripts and are even hesitant about ascribing motives and objectives to the individuals whose talk is analysed. These scruples seem to be shared with the original ethnomethodological critiques. If on the other hand CA is to be seen as part of a broader social science then whilst it might still be admitted that these larger issues (Goffman's 'first' questions) cannot be studied directly, they should not be regarded as radically distinct from the concerns of CA. Three forms of argument can begin to show how CA is to be assimilated.

The first of these is that, in practice, CA does not rigorously separate conversational rules from other, more broadly social rules. Studies in CA suggest that some forms of adjacency pairing admit of two responses. Hence assessments can be assented to or rejected. And within these options one is said to be preferred whilst the other is dispreferred. An acceptance is the preferred pairing to an invitation. Refusals are usually given only with hesitation and are often 'marked' or embellished in some way. Equally, it is proposed, assents are the preferred responses to assessments. However, people relatively frequently make self-denigratory statements like 'I'm awfully dull to talk to'. In this case, as Levinson has recently commented,[26]

> By the preference for agreements after assessments, if A self-denigrates, an agreement from B is preferred. But by an independent principle of a different order, namely a norm enjoining the avoidance of criticism, B should avoid such agreement. The latter principle in fact generally takes precedence.

The 'norm' about avoiding criticism can hardly be a rule of conversational structure since it is about 'the avoidance of *criticism*' not about the location of the speech tokens called 'criticisms'. It is indeed a norm of 'a different order'. Once supplementary conceptions of this kind are invoked then there is no longer any methodological haven for CA away from the customary concerns of sociology.

The second issue concerns the explanations which are given of the conversational structures themselves. In their study of courtroom interaction, Atkinson and Drew relate the formalised organisation of talk to the functional requirements of the judicial proceedings.[27] They argue that the technical demands of making law and passing practical judgement encourage the evolution of certain interactive patterns. Now, while this argument is quite plausible and can readily be extended to account for the prevalence of turn-taking machinery and so on in the wider context, this dependence on a functionalist stance is seldom made explicit. Such a stance does, however, appear to be the only means which exponents of CA adopt for presenting their findings in a unified fashion. In so far as conversational communication is an impersonal, biological development then it would appear to be properly treated by evolutionary arguments. But where rules of conversation, as in the courtroom, have not sprung directly from nature it is inappropriate to assume that they have come about for purely technical reasons. Such a methodology could threaten to make all institutions of interaction into 'the best possible arrangements in the best of all possible worlds'.

Finally, CA might claim a non-naturalistic status by virtue of being a branch of linguistics rather than part of social science. On this view, as was described in the discussion of symbol systems in Chapter 1, if some part of social inquiry was concerned only with the disclosure and clarification of meanings it could, like linguistics, be non-naturalistic. However, this argument cannot be sustained in the case of CA. In the last chapter it was maintained that ethnomethodological studies inevitably make the claim to be able to recognise various accounts as being of a similar or related type. Similarly, exponents of CA must be able to recognise, for instance, certain utterances as compliments and others as greetings. In general terms these categories have to be understood as types of speech act and consequently depend on a naturalistic imputation of motives, wants and beliefs. Advocates of CA maintain, on the contrary, that they do not impute such entities in particular cases - instead they are concerned only with the formal properties of such utterance-types. For example, CA is concerned with the form of compliments whether they are used flatteringly or to damn with faint praise. Accordingly, it is maintained that naturalistic assertions about people's individual motives are beside the point. Whilst this defence has a certain plausibility in individual instances it cannot work as a general procedure. For one thing, people routinely do make judgements of motives and so on resulting from others' speech acts so a naturalistic attitude is possible. And, more significantly, unless there is, so to speak, something fundamentally complimentary about compliments, there is no basis for the categorisation at all. In the last instance, the categorisations themselves contain either the implicit claim that the primary meaning of complimentary utterances is a compliment or a statistical assertion that utterances identified as compliments are frequently used in a complimentary way. Although the practice of CA may successfully avoid ascriptions of individual motives and wants, the implicit epistemological claim is that the categories do correspond to some actors' wants and beliefs and to the actions they undertake through speech. Even if the practice of CA appears non-naturalistic, the model of the actor it implies is a thoroughly naturalistic one.

On the interpretation developed here, and despite its avowed claims, ethnomethodological work is not anti-naturalistic when assessed from a post-empiricist position. Pollner's investigations serve to show how precisely everyday reasoning coincides with the analysis of scientific thought developed by Lakatos, and thereby reinforces the argument of Chapter 4 that the demarcation of common or everyday rationality from scientific rationality is impracticable. For its part, CA adds in an intriguing and exciting way to the model of rational action which can be compiled under post-empiricism. Its contribution is analytically similar to that of Goffman although it benefits from being more systematic in the presentation of the conversational rules which comprise the actor's subconscious knowledge, and in the description of the local interpretative contexts created in interaction. Moreover, by stressing how deeply linguistic social interaction is, it avoids the simple dependence on language which can characterise symbolic interactionism. Similarly, in its use of the analyses performed by co-conversationalists CA emphasises how analytic, or in Goffman's terms reflexive, social actors are, and provides a valuable methodological heuristic for checking the analyst's interpretations.[28] However, both in the dependence on the perpetual correctness of actors' judgements and on a straightforwardly functionalist interpretation of conversational structures, the sociology which CA necessarily presupposes can readily become overly benign, even naive. Because it does not occupy the haven which ethnomethodological theory suggests it should, CA cannot rest content with the simple functionalism which insularity might permit.

Dealing with belief

Although not representative of such a coherent trend of empirical work as CA and symbolic interactionism, the sociology of knowledge is discussed here because it continues to be practised by sociologists in a wide variety of specialisms. Indeed, as well as the perennial areas of the sociology of religion and of political beliefs or ideologies it is increasingly applied to scientific and technical belief and to everyday thought. Within the compass of a brief review the particular methodological problems raised by these separate studies clearly cannot be discussed but there will be room to make some procedural points.

The sociology of knowledge has always been confronted with the problem of relativism: if, for example, it is proposed that a belief in the right to own property is in some way a product of specific social conditions and not therefore universally valid, is not the idea of the social determination of knowledge also a social product and consequently of restricted validity? Early ventures in the sociology of knowledge appear to have been fuelled by the notion that some ideas must be correct and that sociological ones are likely to be amongst the correct ones. [29] More recently post-empiricism appears to offer a more systematic solution.

The acceptance of rational beliefs, based on appropriately scientific methods of belief-assessment, can be understood as a straightforwardly rational action while persistent belief in non-scientific ideas must be attributable to some form of external, social force. However, this attractive formulation collapses with Lakatos' inability to specify precisely what is rational about scientific beliefs and exactly when formerly-scientific beliefs cease to be scientific. Moreover, it is not particularly clear what is to be understood by the supplementary notion that social forces cause or lead to beliefs. In some sophisticated presentations, such as the interest theory outlined in Chapter 3, it is proposed that someone's class or social position provides them with a reason for seeing the world in a certain light and that they then select beliefs which are compatible with that social perspective. They choose beliefs which are rational with respect to their social location. But even this account leaves a crucial ambiguity at a methodological level for any particular beliefs that an individual may seem to hold can only have an interpretative connection to their social location. Thus a belief in the genetic inheritance of intelligence is now widely regarded as reactionary whereas in the late nineteenth century it was associated with planned socialist reform. Given that more or less any belief could be linked to almost any social position and that the analyst typically wants to assert that the social production of knowledge goes on unbeknown to the subject (i.e. that the subject regards their beliefs as factual), then the analyst has a methodological *carte blanche* for making attributions of ideological reasons for belief.

Without such a theory, however, the situation is worse still. Given some belief 'B' for which apparently no rational support exists, the analyst asserts that some social cause must be responsible. Yet unless some ulterior reasons are given which make 'B' rational for other purposes (as in interest theory) we have no rationally comprehensible account or 'action story-line' through which to make sense of someone's belief in 'B'. Now, it has already been argued that the regulatory principle of rationality means that social science cannot offer any general explanation of an action/belief for which no combination of wants and background beliefs would give reasonable grounds. Accordingly, such 'explanations' of irrational belief are all form and no content.

Whilst Lakatos' theory of scientific conduct offers a naturalistic account of rational belief and whilst abstract naturalist arguments suggest that a general sociology of knowledge should be practicable, the pragmatic and methodological objections raised above mean that this branch of sociology readily becomes a prejudicial science. It offers to condone certain beliefs as rational whilst purporting to show that others are social products and therefore unwarranted - even though nothing but the vaguest account of their production can be given. Ironically for social science the question of belief remains vital. For one thing, as Pollner's study of the insect-ridden factory demonstrated, it is part of people's everyday reasoning to engage in the sociology of knowledge. The workers conceived the inspector's beliefs as motivated by some ulterior interest and vice versa. Secondly, the centrality of beliefs to the post-empiricist model of action means that questions of belief and knowledge cannot be dispensed with.

Various responses can be offered at this point. It may be the case that further conceptual analysis of rationality and the philosophy of knowledge will allow more rigorous definitions of rational and scientific knowledge, although I find this doubtful. Or through the kinds of extension to the model of action supplied by symbolic interactionism and the form of co-conversant analysis provided by CA, it may prove possible to be more certain of the interpretative connections that actors make between their social context and their knowledge. But in the short run an almost anti-naturalistic approach can be adopted. In many studies of everyday reasoning, analysts simpy re-present the accounts offered by the subjects they study. For example in her study of the role, beliefs and position of housewives Oakley supplies the testimonies of wives whom she interviewed. Before thorough analyses of these accounts are available we may have to make do with the kind of phenomenological understanding of their knowledge and beliefs which the provision of lengthy quotations offers.[30]

Conclusion

In a chapter of this sort it is clearly impossible to pass complete and rounded judgements on the kinds of sociological work discussed, still less to imply that the whole of empirical sociology has been scrutinised. However, I hope to have show how a post-'post-empiricist' analysis of social science can offer practical verdicts and to have drawn attention to the kinds of methodological issues which must be confronted in sociological practice and which count in the evaluation of social research. I also hope to have offered some substantive analyses.

The general theme of these reviews has been the ironic obligation to offer accounts of social action – even sometimes in the form of statistics with an implied 'story-line' – whilst acknowledging all along that the static ideal of natural order place these accounts in an epistemologically anomalous position. One symptom of this precarious situation has been the tendency for competing research programmes – like Weberianism and Marxism – to develop inexorably at the level of accounts. More positively, it has been my suggestion that, far from being mere elements of microsociology, CA and symbolic interactionism indicate that small-scale studies of action, interaction and sociolinguistics contribute directly to the elaboration of the model of action and thereby clarify the basis on which 'macro'-accounts inevitably rest. Studies of this sort develop the theory of action by adding to the instrumentally rational model of action embodied in Lakatos' epistemological theory of conduct. In many areas, therefore, the extension of social knowledge depends on the elaboration of the model of action which only small-scale studies of this sort can offer.

CHAPTER 6

Conclusion: philosophy, method and practice

Science and social science

The present study occupies a position in a long tradition of inquiry into the scientific standing of sociology. The sheer number of the attempts to answer this question must indicate that, to say the least, their failure rate in the past has been high. Some comment is thus called for on the worth of this inquiry and on my presumption that the verdicts I have offered may be of practical utility. The foremost danger is that any attempt to determine whether some discipline is scientific will simply degenerate into a semantic exercise on the meaning of 'scientific'. Throughout this study I have sought to avoid merely semantic issues, principally by referring to a broad range of sciences (including the biological and geological along with the customary physical sciences) and by including empirical studies of scientific practice as recorded by historians and sociologists. But supposing that semantic quibbles have been avoided, why precisely should it matter that social science can be deemed scientific? It can only matter if the answer is more than the completion of an epistemological labelling exercise; particularly, if the answer helps to resolve practical, procedural doubts. There are consequently several sorts of reasons why it matters: firstly, because there are specific 'scientific' techniques (like statistical modelling) whose application to social phenomena is contentious. Secondly, social scientists use the epithets 'scientific' and 'pseudo-scientific' as a major part of the ritual language of denunciation. It is also likely that issues in the philosophy of science - such as the nature of causation - will overlap with substantive themes of the philosophy of social science. Each of these issues could be addressed in isolation from the original question, and each one is of more practical significance than the more general question alone. However, the question about naturalism provides a good vehicle for addressing them in an orderly way and holds out the prospect of supplying coherent and mutually supporting answers to all the separate procedural issues.

In many respects my conclusions show how artifical the general query about the scientific status of sociology is. For although I have argued that sociology is naturalistic, the same arguments which indicate this also show that academic disciplines like history and major aspects of everyday reasoning are equally naturalistic. Being naturalistic is not therefore a sufficient entitlement for the honorific label of a science and it would indeed be semantic to argue the point about 'scientificness' much further. Since the formal attempts at demarcation, as advanced by Lakatos and Popper, do not succeed either there appears to be no point at which the limit to science must be set.[1] This is not to say that the question of naturalism in sociology is fruitless. As has been shown it is important to be able to align sociology with the naturalistic sciences in contradistinction to the non-naturalistic methodologies of, for example, art criticism and certain aspects of linguistics. It is apparent – at least at a level of epistemological principle – that social science can aim at explanations of action and not just at explications of social meanings.

While my conclusions about the epistemologically naturalistic status of social science do therefore indicate that in the most general terms we can have confidence in attempts to extend sociological knowledge and to offer social scientific explanations, these conclusions do not guarantee that sociology can have the kind of success normally associated with natural science. For one thing, a naturalistic status says nothing about the ease with which knowledge can be developed: a biological theory of brain intelligence is epistemologically on the cards but far from successful as yet. There are additional difficulties confronting social science and it has been argued that in one way or another these centre on the role of rationality in sociological explanations. These difficulties have attracted the competing attentions of ethnomethodologists on the one hand, and Lakatosians and rational-actor theorists on the other. And while neither group is consistently correct, their conflicting views allow the methodological standing of the social sciences to be assessed. Both sets of analysts agree that human action is explained by the imputation of wants and beliefs and of a decision-making mechanism: individual's actions are made sense of through the discovery of a rational connection between those actions and the individual's imputed motives, desires and knowledge. Moreover, it is supposed that the individual has also used this same rational assessment in moving from their own wants and beliefs to the actions, albeit perhaps subconsciously or unconsciously. Where one set of theorists seeks to raise this explanatory scheme to the foundation of a science through the iterative investigation of wants and beliefs, ethnomethodologists maintain that this scheme is a mere formalisation of the 'documentary method'. On this latter view, any imputations of beliefs and desires are incorrigibly contextual, indexical and uncertain. References to such inner properties usually occur in the course of giving 'accounts' and are only *ex post facto* rationalisations of action. The claim of this study has been that while the former view is epistemologically superior, the ethnomethodological viewpoint is often a better indication of the practical situation of social science.

More specifically, this judgement has been reached because rational decision theory is methodologically precarious since a degree of rationality is presupposed in the interpretation of behaviour. It is also empirically precarious since people's wants are so various and changeable that the attribution of wants and beliefs could never be accomplished with practical certainty. On the other hand, the alternative, ethnomethodological programme is exaggerated since even if the actual array of wants and beliefs is beyond depiction at any particular time, this does not indicate that such factors do not explain action. Admittedly these factors represent an explanatory ideal, unrealistically posited in specific instances, but in principle they offer properly naturalistic explanations. Social science is thus naturalistic but in a methodologically unsatisfactory sense for its explanatory variables remain crucially ambiguous and uncertain. Sociology is fated perpetually to offer explanatory accounts or, in Goldthorpe's terms, action story-lines, but accounts which are founded on demonstrably frail suppositions. And this frailty is not just the danger of falsifications which confronts all empirical science but a perpetual methodological anxiety. In response to this methodological impasse, I have proposed that the inevitability of offering accounts does not mean, as ethnomethodologists would wish to suggest, that such activity should be avoided by the analyst. It should not be since the knowledge claims produced at worst cannot be epistemologically inferior to common sense. Moreover, in so far as ethnomethodology itself depends on the interpretation of accounts as situationally appropriate rationalisations, imputed motives and wants (to do with face-saving and so on) are central to its procedures also. More analytically beneficial, however, would be the extension and elaboration of the model of the actor in sociology which can be accomplished through strategies similar to those of CA, symbolic interactionism and studies of scientific reasoning.

Such a methodological middle course, while rather pluralistic, does offer a distinctive assessment of a number of methodological mainstays of social science. A number of studies, particularly of macrosociological concepts, acknowledge that there is no single operationalisation of certain concepts. Calvert's examination of class offers a recent example of this practice. Finding that all precise meanings and operationalisations of class are contentious and methodologically imprecise, he adopts Gallie's proposal that some leading concepts of social science are 'essentially contested concepts'. He quotes Gallie's claim that '[class is one of a family of concepts] the proper use of which inevitably involves endless disputes about their proper uses on the part of their users'.[2] To qualify as an essentially contested concept, Gallie suggests, a term must be evaluative, complex, mutable, open to a number of initial interpretations, and used in defensive and aggressive arguments. Thus, it is implied that such concepts are constitutive of, or closely allied with, political debate. And it is this political utility which is regarded as causing the uncertainty and variability of meaning which characterises such concepts. The implicit claim is that such concepts do not have a 'scientific' stability of meaning because they occupy a special place in political, evaluative argumentation. However, this claim which depends on presenting these concepts as significantly special or peculiar can be seen to be representative of the predicament of social scientific concepts *per se*. Authors who invoke this idea of a special kind of concept do not clarify in advance why some sub-set of concept should be essentially contested; how then can one tell which concepts are disputed because they are essentially contested and which concepts have simply not been decided on? According to the alternative interpretation put forward in the last two chapters, it would be precisely the indexical, contextual character of sociological descriptions and the indeterminate operationalisation of sociological concepts – which arises from people's own interpretative skills – which lead to contention. Rather than reserve 'essentially contested' status for certain recalcitrant concepts, it should be seen as a general characteristic of sociological descriptions.

None the less, the extensive resources of practical reasoning, as documented by Pollner but equally well adumbrated by Lakatos, ensure that everyday life is not riddled with essential contests over the application or meaning of words. A wealth of contextual features can be called on in the course of smoothing over any apparent epistemological conflicts. As in Pollner's traffic court, such difficulties can be transmuted into disputes over matters of apparent fact. And this tendency leads to an issue almost the polar opposite of Gallie's problem: the realism of common-sense descriptions. Throughout this study post-empiricism has chiefly been identified with conventionalism, although the realist view has been discussed briefly. Something of a formal paradox seems to arise here for I have argued both that everyday reasoning is naturalistic and that it is predominantly realistic, whereas philosophical accounts of naturalism are primarily conventionalist, not realist. There is a two-part resolution to this paradox which can be quickly outlined. Firstly, philosophical conventionalism is in many ways unstable, as suggested in Chapter 4, since there is ultimately no way of demonstrating that the Lakatosian (or any other) method is good unless it subserves some higher purpose - and that higher purpose is usually disclosing more about the world. Thus no matter how refined the conventionalism, it frequently contains some implicit dependence on realism. Accordingly there is nothing curious about the attitude of everyday reasoning being ultimately realistic. The second element is that, with a static model of natural order and unchanging hard core as I have claimed to detect in social science, conventionalism becomes in any case practically indistinguishable from realism. A convention to which there exist no alternatives can readily be treated as if it were a direct presentation of reality.

To summarise then: this inquiry into naturalism and social science has touched on all three types of reasons why the question of sociology's scientific standing matters. And it has done so not abstractly but in the light of the practical issues involved. It has looked at 'science' as a term of denunciation and, for example, found that the ethnomethodologists' use of the term is based on a misapprehension about scientific knowledge. Their favoured opposition between literal and indexical descriptions does not stand up, neither does their strict separation of ethnomethodology from avowedly 'scientific' sociology. Naturalism has also been examined as a licence for the application of scientific techniques to sociology. This licence has been accepted, particularly in the case of statistics, but the practical obstacles to successful operationalisation make this licence far less significant than might have been anticipated. Lastly, the philosophy of science has thrown considerable light on social science, not least in identifying the methodological role or rationality in sociological interpretation and in undermining the (Winchian) idea that there may exist alternative rationalities. This assault on a long-standing abstract question has thus produced valuable pay-offs. It has also brought one particular set of problems to light which are to do with the location of social science on the supposed boundary between empirical science and philosophy, a situation brought about by the unchanging ideal of natural order. It is to this set of issues that I must now turn.

Language and the a priori in social science

The location of social science on the apparent boundary of science in this way allows one to re-examine some alternative models for social knowledge which were introduced earlier. Of the various candidates identified in this study, everyday knowledge seemed to have strong attractions. People who are ordinarily concerned with specific areas of social life clearly have deep and subtle knowledge of their local environment. Yet, when the post-empiricist model of the social actor is expanded to incorporate such features as the interactive competences described by CA and the reflexive understanding identified by Goffman, it becomes apparent that the kinds of knowledge associated with everyday activities do not pose a true alternative. They too are naturalistic in orientation.

A second candidate is philosophy. Its case was most strongly put by Winch arguing from presuppositions derived from Wittgenstein. Allegedly, social science like philosophy was concerned with conceptual analysis and the elucidation of language-games. In the case of the philosophy of science at least this argument has been overtaken by circumstances. With Popper and Lakatos the philosophy of science has been transmuted into a theory of scientific action. And the goal of such work is, at least in part, an empirically applicable account of behaviour. There are, as has been shown, considerable difficulties in the precise empirical realisation of this philosophical account but it seems undeniable that philosophical analyses have become empirical and thus also naturalistic in orientation. Ironically this turn towards empirical evidence contrasts with the obstacles to empiricism which this philosophy of science erected in the first place. By demonstrating how far factual evidence depended on theoretical pre-interpretation this philosophy of science threatened to relegate empirical evidence to a very minor role and perhaps - at the hands of Feyerabend - to banish it altogether. Post-empiricism still awaits the resolution of this problem but its widespread nature indicates that, as intendedly empirical disciplines, social science, natural science and the philosophy of science have great methodological similarities. Even if it is not precisely clear how it is that sociology generates empirical knowledge (and of course the sociology of science suggests many complications in any particular case of knowledge generation) it certainly makes sense to see social science as properly empirical.

Language offered a third alternative, as a system of signification which does not depend on naturalistic understanding. In many respects language is a stronger candidate because, as was suggested in Chapter 1, it appears to afford a different form of comprehension. Instead of aiming to understand how or why the system of signification works, one is simply interested in understanding what is signified. However the researches associated with CA indicate that in practical situations interactants are concerned to assess each other's utterances in ways which draw on contextual information in order to understand the utterance as arising from the person's aims, purposes and so on. Indeed, this view of interactants theorising about each other is central to the model of the actor in all interactionist and ethnomethodological work. Accordingly naturally-occurring, social uses of language have a crucially naturalistic component which accommodates language use to the view of social science being developed in this study. Some linguists might wish to suggest that language does constitute a non-naturalistic system of signification and that the naturalistic concern of interactants with the 'background' to utterances is a derivative form. Even if this is correct, any interactive use of language, whether as direct signification or as a less direct speech form, would still need to be understood in contextually relevant and appropriate ways: even the purest significations would need to be understood, in Goffman's terms, reflexively.

These considerations on language also touch on another anti-naturalist objection which was introduced in Chapter 1. There it was suggested that in everyday interaction people are concerned with each other's intentions in acting and not their (causally relevant) reasons for acting. That is, people are generally interested in identifying what others are doing (for example, repairing the car or going to church) rather than in why they are doing it. The argument was that this form of comprehension is alien to the concerns of natural science. Through the notion of reflexive interpretation it can be seen that, in practical contexts, the understanding of intentions is closely allied with imputations of reasons. As in the case of the seating arrangement in the railway carriage, discussed earlier, estimations of people's intentions are formed against the background of the likely reasons for action ascribed to them. Thus the logical standing of this anti-naturalist argument is close to that of the last point about language as a symbol system. While an abstract case could be made out for linguistic codes or intentions as non-naturalistic forms of comprehension, any practical use of them depends on naturalistic assumptions about people's reasoning. And the necessity of such naturalistic assumptions deprives intentional descriptions and linguistic codes of any distinctively anti-naturalistic status.

Whilst these three putatively non-scientific models for social knowledge fail to demonstrate that, epistemologically speaking, sociology should not be allied with the sciences, each of them indicates a pragmatic way in which social science stands close to the boundary of empirical scientific knowledge. Thus, because people can possess very developed interactional skills and very intense local social knowledge, concerning others' likely aims and beliefs, sociology may seem to concern itself solely with what people know already and therefore, in a limited sense, to be a form of a priori knowledge. Similarly, it is close to the boundary because the supposed fundamental attributes of actors - which make up their philosophical anthropology - are often discussed in philosophical, metaphysical terms, particularly as regards freedom of the will and individuals' rights. Finally, several analysts have seen great significance in the dependence of social knowledge on language. Habermas, for example, has asserted that:[3]

> What raises us out of nature is the only thing whose nature we can know: *language*. Through its structure, autonomy and responsibility are posited for use.

On the other hand, Strawson in his exercise in descriptive metaphysics argues that there are complex, possibly competing metaphysics enshrined in our linguistic practices. These commitments, he suggests, are extremely enduring:[4]

> For there is a massive central core of human thinking which has no history – or none recorded in histories of thought; there are categories and concepts which, in their most fundamental character, change not at all. . . . They are the commonplaces of the least refined thinking; and yet are the indispensable core of the conceptual equipment of the most sophisticated human beings.

No doubt language does in some manner set the perimeter for our thought but whether this is a constraint or a liberation cannot successfully be decided before we know much more about how language is used and understood in interaction. Although social science connects intimately with the a priori in a number of ways this by no means indicates that its naturalistic status is impaired.

One last issue which must briefly be taken up here is the question of values. As was noted in Chapter 1 the arguments of critical theorists and also of some liberal philosophers suggested that social knowledge had an inescapable evaluative component. Because of the closeness of social science to our everyday judgements of morality and values, and because of the proximity of the model of the actor to metaphysics, this is no doubt the case. However the connection does not appear to be a straightforward or systematic one. There is nothing, for example, in the very fluid notion of the rational actor that assists directly in judgements of an evaluative or moral nature nor anything about the social use of language which unambiguously postulates any particular set of values. Moreover, the recognition that social science is naturalistic or science-like in orientation does not have immediate consequences for the issue of values. As is increasingly the case with animal biology, a naturalistic inquiry can raise profound evaluative questions without supplying an immanent solution to them. At the present stage, with an unchanging but profoundly ambiguous concept of the rational actor and a sociology riddled with essentially contested concepts, the evaluative connections are immense but very far from being agreed in detail. Sociology's pragmatic proximity to the a priori definitely does not mean that it is in a position to adopt the ethical or moral mantle of metaphysics no matter how frequently it generates questions of values.

Prognoses

In this final section the arguments presented in this study will be drawn together not simply to summarise the present epistemological position of social science but especially to indicate some of the directions in which it might be developed given the limitations it faces. Although many of the considerations which have been introduced could be construed as criticisms of social science, my conclusion has been that sociology can and should offer explanations of social action and social regularities. And these explanations should be of the kind that post-empiricism would endorse. Thus I agree neither with ethnomethodologists nor with supporters of Winch about the prospects of social science. However, I anticipate that the kinds of explanatory account which sociology generally offers will often appear predictable and unexciting and that competing interpretations will continue to be advanced. Thus the sense of continued breakthrough and consensual agreement which is often (but exaggeratedly) ascribed to the natural sciences will not come to characterise social science - even if all its practitioners sought to follow a post-empiricist, Lakatosian methodology. The so-called Founding Fathers will continue to preoccupy the sociological imagination not because of their (undoubted) intellectual stature but because they offer fundamental explanatory accounts and basic depictions of the rational actor which can be repeatedly used to make sense of new social phenomena.[5] Work of this sort will and must continue but its methodological problems also will persist.

In certain areas work within the macrosociological tradition stands in need of methodological revision. In order to formulate theories of any significant generality these approaches demand that some device is used to associate patterns of individual action with broader, societal trends. As noted in Chapter 4, in the case of Marxism and functionalism this may amount to applying the regulatory principle of rationality to social institutions rather than to individual agents. These explanatory schematas can then be rationalised by imputing a compatible range of wants and beliefs to the agents. When applied to particular instances this procedure runs into difficulties. For one thing individuals have a more complex and shifting range of wants and beliefs than these philosophical anthropologies customarily allow. Moreover, such macrosociological accounts appear inflexibly deterministic since they situate individuals' actions with respect to societal tendencies. It is, I suggest, out of these conflicts that the complex of difficulties usually dubbed the micro/macro problem arises. In order to inject a sense of the openness of action without denying its regularity it is necessary to allow a more diverse, mutable and above all reflexive conception of the actor *and* to accept that the apparent freedom and indeterminacy of social action arises from the local and interactive concerns of actors. The regularities of social life are built upon a negotiated order and these negotiations, while composed of restricted sets of strategies, are as both game theory and CA assert, indeterminate. Hence the kind of research which is most likely to be rewarding at this level will incorporate Goffmanesque studies of reflexive intersubjective understanding, CA-type studies of local interactive expectations, and game-theoretic analyses of negotiations.

In at least the first two of these three areas language and linguistic interaction is central but language must not be permitted to become an infatuation no matter how great its importance for present analytic needs. It must not become an infatuation, as one might say, factually in that not all interactions are even mediated linguistically nor are all our thoughts and emotions couched in language. Neither should language become methodologically overwhelming: no matter how vital linguistic and para-linguistic features are in governing interaction and in establishing local interpretative contexts, not all social actions are acts of communication or even of exchange. Other forms of social activity, corresponding to the questions which Goffman described as 'first', sometimes consist in individuals disregarding or overlooking local interpretational obligations. Such forms of activity can only be said to relate to social power; but power is not as readily subjected to analysis as are linguistic forces. Its description must, however, remain on the agenda.[6] Equally, in some contexts at least, say of confinement or subjugation, what people do is very narrowly constrained no matter what their individual constellations of wants and beliefs may be and no matter how exactingly they exercise the small space which remains for negotiation. Such contexts of unfreedom must not be excluded from study either.

The other direction for the extension of sociology which I anticipate is towards a methodological deconstruction of the individual. So long as we identify the ideal of natural order of the rational actor with particular individuals, sociology will offer recognisable but predictable and contested explanations. Even with such apparent challenges to the individual as are posed by Freudianism, the person is simply resolved for analytic purposes into an assemblage of three rational actors. But hints arising from Strawson's descriptive metaphysics and from sociolinguistics and discourse analysis suggest that each individual's identity is constituted of alternative linguistic registers. Furthermore, more diffuse claims from feminism, structuralist psychoanalysis and literary analysis propose in various ways that the self is multi-faceted and confict-ridden.[7] While for the present, therefore, a greater understanding of the locally-directed, reflexive action of the individual is required, such work may well indicate that one foreseeable direction for new research is away from the idea that the site of rational action is in the individual or self as traditionally understood.

Notes and References

In these notes and references two abbreviations are used: the journals *Social Studies of Science* and *Philosophy of the Social Sciences* are indicated by *SSS* and *PSS* respectively.

Introduction

1 Aspects of this history are discussed in W. Outhwaite, *Understanding Social Life* (London: Allen & Unwin, 1975) and in G. Hawthorn, *Enlightenment and Despair* (Cambridge: CUP, 1976).
2 See R. Bhaskar, *The Possibility of Naturalism* (Brighton: Harvester, 1979); R. Keat and J. Urry, *Social Theory as Science* (London: Routledge & Kegan Paul, 1975), second edn 1982; D. Papineau, *For Science in the Social Sciences* (London: Macmillan, 1978); and D. Thomas, *Naturalism and Social Science* (Cambridge: CUP, 1979).
3 These issues are explored further and documented in Chapter 3 and in the references cited there.

Chapter 1 *Empiricist naturalism and the anti-naturalist response*

1 For a brief history of these considerations see the opening chapters of C. Marsh, *The Survey Method* (London: Allen & Unwin, 1982), pp. 9–47.
2 A lucid but moderate statement of this line of thought is provided in A.J. Ayer, *Language, Truth and Logic* (Harmondsworth: Penguin, 1971), the comparatively late date of which indicates the continuing appeal of this philosophy at many levels, the first edition was in 1936.
3 C.G. Hempel, *Philosophy of Natural Science* (Englewood Cliffs: Prentice-Hall, 1966), and E. Nagel, *The Structure of Science* (London: Routledge & Kegan Paul, 1961).
4 Hempel, *Philosophy of Natural Science*, pp. 2–8.
5 Hempel, *Philosophy of Natural Science*, pp. 47–8.
6 Cited on p. 25 of C. Taylor, 'Neutrality in Political Science', in P. Laslett and W.G. Runciman, *Philosophy, Politics and Society*, III (Oxford: Blackwell, 1969), from V. van Dyke, *Political Science* (London: Stevens; and Stanford: Stanford University Press, 1960) p. 192.

7 See the methodological discussion in R.G. Lipsey, *An Introduction to Positive Economics* (London: Weidenfeld & Nicolson, 1979), pp. 3–17.

8 For a discussion of the meanings of positivism as it has been interpreted in the social sciences see P. Halfpenny, *Positivism and Sociology: Explaining Social Life* (London: Allen & Unwin, 1982).

9 On Durkheim's practical use of 'positivism' see Keat and Urry, *Social Theory as Science*, pp. 83–7.

10 C.F.A. Pantin, *The Relations Between the Sciences* (Cambridge: CUP, 1968) provides one of the only systematic studies of the cognitive inter-dependencies of the sciences.

11 An example famous for its influence on Parsons is R. Bales, *Interaction Process Analysis* (Cambridge, Mass.: Addison-Wesley, 1950). The social-psychological literature is full of further examples dealing, for instance, with social learning, inter-personal perception and group behaviour.

12 For an analysis of science as a learning machine and a consideration of the effects of an unstable environment see M.B. Hesse, 'In Defence of Objectivity', *Proceedings of the British Academy*, 58 (1973), pp. 275–92.

13 On speech acts see J.R. Searle, *Speech Acts* (Cambridge: CUP, 1969) and the discussion in S.C. Levinson, *Pragmatics* (Cambridge: CUP, 1983), pp. 226–83

14 On the notion of competence see N. Chomsky, *Topics in the Theory of Generative Grammar* (The Hague and Paris: Mouton, 1966), pp. 9–11.

15 A good introduction to this literature is afforded by E.R. Leach, *Culture and Communication* (Cambridge: CUP, 1976).

16 C. Taylor, *The Explanation of Behaviour* (London: Routledge & Kegan Paul, 1964), p. 33.

17 See J. Habermas, *Knowledge and Human Interests* (London: Heinemann, 1972), and his 'A Postscript to *Knowledge and Human Interests*', *PSS*, 3 (1973), pp. 157–89, as well as K.O. Apel, 'Communication and the Foundations of the Humanities', *Acta Sociologica*, 15 (1972), pp. 7–27. A comprehensive review of these arguments is supplied in T. McCarthy, *The Critical Theory of Jurgen Habermas* (London: Hutchinson, 1978), pp. 53–125.

18 This particular issue can be explored further through Chapter 9 of Habermas, *Knowledge and Human Interests*, pp. 191–213, and through G. Hawthorn, 'The Interest of Scientific Knowledge', *Science Studies*, 3 (1973), pp. 78–87.

19 P. Winch, *The Idea of a Social Science and its Relation to Philosophy* (London: Routledge & Kegan Paul, 1958).

20 Winch, *The Idea of a Social Science*, p. 89.

21 A. MacIntyre, 'The Idea of a Social Science', in B.R. Wilson (ed.), *Rationality* (Oxford: Blackwell, 1970), p. 113.

22 P. Winch, 'Understanding a Primitive Society', in Wilson, (ed.) *Rationality*, p. 95.

23 On the ancestry of *Verstehen* see Outhwaite, *Understanding Social Life*, pp. 11–37.

24 See M. Horkheimer, *Critical Theory* (New York: Seabury, 1972) and the studies collected in P. Connerton (ed.), *Critical Sociology* (Harmondsworth: Penguin, 1976).

25 Taylor, 'Neutrality in Political Science' in Laslett and Runciman, *Philosophy, Politics and Society*, III, pp. 25–57.
26 P.F. Strawson, *Freedom and Resentment and Other Essays* (London: Methuen, 1974), pp. 1-25.
27 For the principal instances see H. Garfinkel, *Studies in Ethnomethodology* (Englewood Cliffs: Prentice-Hall, 1967), pp. 35–75. Additional discussions can be found in A.V. Cicourel, *Method and Measurement in Sociology* (New York: Free Press, 1964); D. Benson and J.A. Hughes, *The Perspective of Ethnomethodology* (London: Longman, 1983); and W. Handel, *Ethnomethodology: How People Make Sense* (Englewood Cliffs: Prentice-Hall, 1982).

Chapter 2 *The naturalism of the new philosophy of science*

1 See the works cited in note 2 to the Introduction and, on science more generally, A.F. Chalmers, *What is this Thing called Science?* (Atlantic Highlands: Humanities Press, 1976; and Milton Keynes: Open University Press, 1978, revised edn 1982); R. Harré, *The Philosophies of Science* (Oxford: OUP, 1976); J. Losee, *A Historical Introduction to the Philosophy of Science* (Oxford: OUP, 1980); and R. Bhaskar, *A Realist Theory of Science*, second edition (Brighton: Harvester, 1978).
2 On such definitional or (as they are sometimes called) 'analytic' statements see W.V.O. Quine, 'Two Dogmas of Empiricism', in his *From a Logical Point of View* (New York: Harper & Row, 1963).
3 P.K. Feyerabend, *Against Method* (London: New Left Books, 1975).
4 The major statements of this methodology are to be found in K.R. Popper, *The Logic of Scientific Discovery*, reprint edn (London: Hutchinson, 1980) and his *Conjectures and Refutations* (London: Routledge & Kegan Paul, 1963). A useful overview is provided in Chalmers, *What is this Thing called Science*, pp. 38–49.
5 T.S. Kuhn, *The Structure of Scientific Revolutions* (Chicago and London: University of Chicago Press, 1970), enlarged from the edition of 1962.
6 A valuable collection of papers on this issue is available in I. Lakatos and A. Musgrave (eds.), *Criticism and the Growth of Knowledge* (Cambridge: CUP, 1970). The great majority of Lakatos' relevant work is contained in his *Philosophical Papers*, 2 vols. (Cambridge: CUP, 1978), ed. J. Worrall and G. Currie, especially vol. I, *The Methodology of Scientific Research Programmes.*
7 It should be recalled that these anomalies do not have a direct, factual status; they are still relative to the presuppositions of the research programme.
8 On Newton's alchemy see B.J.T. Dobbs, *The Foundation of Newton's Alchemy: or the Hunting of the Greene Lyon* (Cambridge: CUP, 1975).
9 See Bhaskar, *A Realist Theory of Science*, and R. Harré, *The Principles of Scientific Thinking* (London: Macmillan, 1970).
10 These issues are clearly brought out in B. Barnes and J. Law, 'Whatever Should be done with Indexical Expressions?', *Theory and Society*, 3 (1976), pp. 223-37. As these authors suggest, indexicality is nicely illustrated by the case studies in I. Lakatos, *Proofs and Refutations* (Cambridge: CUP, 1976).

11 This argument is carefully expounded in Papineau, *For Science in the Social Sciences* (London: Macmillan, 1978), pp. 78–90.
12 An extensive yet introductory discussion of the rationality of human action is to be found in A. Heath, *Rational Choice and Social Exchange* (Cambridge: CUP, 1976).
13 See for example R.D. Laing, *Self and Others* (London: Tavistock, 1969), pp. 35–7.
14 P. Abrams, *Historical Sociology* (Shepton Mallet: Open Books, 1982), pp. 1–17.
15 Among the most thorough critiques of functionalism are A. Giddens, 'Functionalism: *après la lutte*', in his *Studies in Social and Political Theory* (London: Hutchinson, 1977), pp. 96–129; his *Central Problems in Social Theory* (London: Macmillan, 1979), pp. 96–130; and J. Elster, *Ulysses and the Sirens: Studies in Rationality and Irrationality* (Cambridge: CUP, 1979), pp. 28–35.
16 On this aspect of Marxism see Chapter 1 of J.E. Roemer, *Analytical Foundations of Marxian Economic Theory* (Cambridge: CUP, 1981) and, for the cunning of history, Hawthorn, *Enlightenment and Despair*, pp. 42–50.
17 These issues are nicely illustrated in M. Hollis, 'The Limits of Irrationality' and 'Reason and Ritual', both in B.R. Wilson (ed.), *Rationality* (Oxford: Blackwell, 1970), pp. 214–20 and 221–39.
18 The similarity between arguments for alternative rationalities and for the existence of ethno-methods is indicated by the welcome extended to exotic anthropologies by some ethnomethodologists: see H. Mehan and H. Wood, *The Reality of Ethnomethodology* (New York and London: Wiley, 1975), pp. 27–31.

Chapter 3 *Studies in the practice of science*

1 Philosophers' examples include Lakatos, *Proofs and Refutations* (Cambridge: CUP, 1976), and H. Frankel, 'The Career of Continental Drift Theory: An Application of Imre Lakatos' Analysis of Scientific Growth to the Rise of Drift Theory', *Studies in the History and Philosophy of Science*, 10 (1979), pp. 21–65. These can be contrasted with the studies collected in H.M. Collins (ed.), *Sociology of Scientific Knowledge: A Sourcebook* (Bath: Bath UP, 1982).
2 See, for example, the study of the history of scientific medicine: P.W.G. Wright, 'A Study in the Legitimisation of Knowledge: The "Success" of Medicine and the "Failure" of Astrology', in R. Wallis (ed.), *On the Margins of Science: The Social Construction of Rejected Knowledge, Sociological Review Monograph* 27 (Keele: University of Keele, 1979), pp. 85–101.
3 M. Mulkay, *Science and the Sociology of Knowledge* (London: Allen & Unwin, 1979), pp. 39–42.
4 Thus B. Barnes and S. Shapin (eds.), *Natural Order: Historical Studies of Scientific Culture* (Beverly Hills and London: SAGE, 1979), is not specifically concerned with naturalism in the present sense. For the connection with Winch, see H.M. Collins, 'An Empirical Relativist Programme in the So-

ciology of Scientific Knowledge', in K.D. Knorr-Cetina and M. Mulkay (eds.), *Science Observed: Perspectives on the Social Study of Science* (Beverly Hills and London: SAGE, 1983), pp. 85–100.

5 See H. Rose and S. Rose (eds.), *The Political Economy of Science* (London: Macmillan, 1976), especially the editors' comments on pp. 14–31; and L. Winner, 'Do Artifacts have Politics?', *Daedalus*, 109 (1980), pp. 121–36.

6 These pressures are examined in J.R. Ravetz, *Scientific Knowledge and its Social Problems* (Oxford: Clarendon, 1971).

7 For a review of the finalisation debate see J. Symes, 'Policy and Maturity in Science', *Social Science Information*, 15 (1975), pp. 337–47.

8 J. Slack, 'Class Struggle Among the Molecules', in T. Pateman (ed.), *Countercourse* (Harmondsworth: Penguin, 1972), pp. 202–17.

9 On ideology see Rose, (eds.), *The Political Economy of Science*; and R.M. Young, 'Evolutionary Biology and Ideology - Then and Now', *Science Studies*, 1 (1971), pp. 177–206.

10 This issue is examined in detail in D. Bloor, *Knowledge and Social Imagery* (London: Routledge & Kegan Paul, 1976).

11 Principal examples of this form of analysis include: H. M. Collins, 'The Seven Sexes: A Study in the Sociology of a Phenomenon, or the Replication of Experiments in Physics', *Sociology*, 9 (1975), pp. 205–24; H.M. Collins, 'Stages in the Empirical Programme of Relativism', *SSS* 11 (1981), pp. 3–10; H.M. Collins, 'The Role of the Core-Set in Modern Science: Social Contingency with Methodological Propriety in Science', *History of Science*, 19 (1981), pp. 6–19; H.M. Collins and T.J. Pinch, 'The Construction of the Paranormal: Nothing Unscientific is Happening', in Wallis, *Margins of Science*, pp. 237–70; and H.M. Collins and T.J. Pinch, *Frames of Meaning: The Social Construction of Extraordinary Science* (London: Routledge & Kegan Paul, 1982).

12 See Collins, 'The Seven Sexes'.

13 H.M. Collins, 'The Function of Calibration in Experimental Science', paper presented to the combined meeting of HSS, PSA, SHOT and 4S, Philadelphia, 28–31 August 1982.

14 Collins and Pinch, 'The Construction of the Paranormal', especially pp. 253–55 and 259–63.

15 See Collins and Pinch, 'The Construction of the Paranormal', p. 244.

16 This theory is set out in B. Barnes, *Interests and the Growth of Knowledge* (London: Routledge & Kegan Paul, 1977); and in S. Shapin, 'The Politics of Observation: Cerebral Anatomy and Social Interests in the Edinburgh Phrenology Disputes', in Wallis, (ed.), *On the Margins of Science*, pp. 139–78.

17 D. MacKenzie, 'Statistical Theory and Social Interests: A Case Study', *SSS* 8 (1978), pp. 35–83.

18 See B. Barnes, 'Natural Rationality: A Neglected Concept in the Social Sciences', *PSS* 6 (1976), pp. 115–26. Similar themes are explored in his *T.S. Kuhn and Social Science* (London: Macmillan, 1982), especially pp. 94–114.

19 See B. Barnes and D. MacKenzie, 'On the role of Interests in Scientific Change', in Wallis, (ed.), *On the Margins of Science*, pp. 53ff.

20 This distinction is introduced in S. Woolgar, 'Laboratory Studies: A Comment on the State of the Art', *SSS* 12 (1982), p. 487.
21 K.D. Knorr-Cetina, 'The Ethnographic Study of Scientific Work: Towards a Constructivist Interpretation of Science', in Knorr-Cetina and Mulkay, *Science Observed*, p. 117.
22 S. Woolgar, 'Interests and Explanation in the Social Study of Science', *SSS* 11 (1981), pp. 365–94.
23 B. Latour and S. Woolgar, *Laboratory Life: The Social Construction of Scientific Facts* (London and Beverly Hills: SAGE, 1979), pp. 151–87.
24 G.N. Gilbert and M. Mulkay, 'Introduction' to their *Opening Pandora's Box: A Sociological Analysis of Scientists' Discourse* (Cambridge: CUP, 1984).
25 See the chapter concerning consensus in Gilbert and Mulkay, *Opening Pandora's Box*.
26 M. Mulkay and G.N. Gilbert, 'Joking Apart: Some Recommendations Concerning the Analysis of Scientific Culture', *SSS* 12 (1982), pp. 585–613.
27 Indeed B. Barnes, in his *Scientific Knowledge and Sociological Theory* (London: Routledge & Kegan Paul, 1974), recommends natural science as the model for sociology: see pp. 78–84.

Chapter 4 *Post-empiricism revisited*

1 For this and other philosophers' criticisms of Lakatos' theory see the discussion in L. Laudan, *Progress and its Problems: Towards a Theory of Scientific Growth* (London: Routledge & Kegan Paul, 1977), pp. 76–120.
2 The way in which historical accounts can be constructed around the Lakatosian framework is examined in M. Mulkay, J. Potter and S. Yearley, 'Why an Analysis of Scientific Discourse is Needed', in Knorr-Cetina and Mulkay, *Science Observed*, pp. 172–82.
3 See Laudan, *Progress and its Problems*, for the problems of discerning the rationality of past scientific actions. These problems are also confronted in B. Barnes, 'Vicissitudes of Belief', *SSS* 9 (1979), pp. 247–63.
4 See, for example, Popper, *Logic of Scientific Discovery* (London: Hutchinson, 1980), pp. 19 and 50.
5 Lakatos, *Philosophical Papers*, I pp. 153–4.
6 For an account of operationalism see Losee, *A Historical Introduction to the Philosophy of Science*, pp. 175–8 and Harré, *The Philosophies of Science* (Oxford: OUP, 1976), pp. 62–80.
7 For the case of economics see Lipsey, *An Introduction to Positive Economics* (London: Weidenfeld & Nicolson, 1979), p. 542, and for psychology see G. Böhme, 'Cognitive Norms, Knowledge-Interests and the Constitution of the Scientific Object', in E. Mendelsohn, P. Weingart and R. Whitley (eds.), *The Social Production of Scientific Knowledge, Sociology of the Sciences Yearbook* I (Dordrecht: Reidel, 1977), pp. 129–41.
8 See T.S. Kuhn, 'Objectivity, Value Judgment, and Theory Choice', in his *The Essential Tension: Selected Studies in Scientific Tradition and Change* (Chicago and London: University of Chicago Press, 1977).

9 M. Mulkay and G.N. Gilbert, 'Putting Philosophy to Work: Karl Popper's Influence on Scientific Practice', *PSS* 11 (1981), pp. 389–407.

10 As has already been mentioned (see notes 2 and 3), attempts to apply these philosophical rules to historical instances will also be problematic for, if the history is in doubt, the interpretation of the rules will also be contestable.

11 Elster, in *Ulysses and the Sirens* (Cambridge: CUP, 1979), pp. 4–28, discusses many techniques for amplifying our understanding of rationality. He also points out that natural systems which simulate rationality (like evolutionarily adaptive mechanisms) are generally not optimally rational. Evolutionary adaptations are only rational for the short term: for example, a predator which evolved an ability to go hungry when it was necessary to conserve its prey would be eliminated in the short term by its less circumspect fellows even though their behaviour might wipe out both the prey and, thereby, the predators in the long run. The long-term rationality of the self-denying behaviour would not be expected to become dominant in this species, even though such behaviour might be advantageous. Such distinctions, between short-term and optimal rationality, are among the results of philosophical analyses of rationality.

12 S.E. Toulmin, *Foresight and Understanding* (London: Hutchinson, 1961). Once again the word 'natural' may lead to confusion: Toulmin's 'Ideals of Natural Order' are not the same as Barnes and Shapin's *Natural Order*.

13 Toulmin, *Foresight and Understanding*, p. 54.

14 D. Smith, 'Theorizing as Ideology' in R. Turner (ed.), *Ethnomethodology* (Harmondsworth: Penguin, 1974) pp. 41–5. For a related discussion concerning attributions of interests in the sociology of science see Woolgar, 'Interests and Explanation', and the reply, B. Barnes, 'On the "Hows" and "Whys" of Cultural Change', *SSS* 11 (1981), pp. 481–98.

15 See Cicourel, *Method and Measurement in Sociology* (New York: Free Press, 1964), pp. 92ff; and Mehan and Wood, *The Reality of Ethnomethodology* (New York and London: Wiley, 1975), pp. 37–73, especially pp. 46–54.

16 P. Feyerabend seeks to show the generality of Lakatos' version of rationality by applying the methodology of scientific research programmes to organised crime, see his 'Consolations for the Specialist' in Lakatos and Musgrave (eds.), *Criticism and the Growth of Knowledge* (Cambridge: CUP, 1970), pp. 197–231, especially pp. 200–1.

17 On the demarcation criterion as formulated by Popper, see the discussion in A. Tudor, *Beyond Empiricism: Philosophy of Science in Sociology* (London: Routledge & Kegan Paul, 1982), pp. 120–7.

18 Lakatos, *Philosophical Papers* I, pp. 3 and 5.

19 In the light of the point raised at note 11, the exact degree of rationality imputed to species may not be the same as for ids and classes, but the methodological principle of imputed rationality is the same.

20 Garfinkel, *Studies in Ethnomethodology* (Englewood Cliffs: Prentice-Hall, 1967), p. 1. Garfinkel also expresses his scepticism about explanations in terms of internal states in his 'A Conception of, and Experiments with,

"Trust" as a Conditions of Stable Concerted Actions', in O.J. Harvey, *Motivation and Social Interaction* (New York: Ronald Press, 1963), p. 190.

21 Representative criticisms can be found in S. Bruce and R. Wallis, 'Rescuing Motives', *British Journal of Sociology*, 34 (1983), pp. 61–71; and in J. Goldthorpe, 'A Revolution in Sociology?', *Sociology*, 7 (1973), pp. 449–62. Less critical judgements are passed in the contributions in G.N. Gilbert and P. Abell (eds.), *Accounts and Action* (Farnborough: Gower, 1983).

Chapter 5 *Science and social research practices*

1 See B.N. Meltzer, J.W. Petras and L.T. Reynolds, *Symbolic Interactionism: Genesis, Varieties and Criticisms* (London: Routledge & Kegan Paul, 1975), pp. 53–67. E. Goffman's work consistently attends to the world of the middle classes, as many of the examples in, say, *The Presentation of Self in Everyday Life* (London: Allen Lane, 1969), testify.

2 This study by H.S. Becker (from the *American Journal of Sociology*, 59 (1953), pp. 41–58) is widely available in the Open University, School and Society Course Team's *School and Society: A Sociological Reader* (London: Routledge & Kegan Paul, 1971) – see 'Becoming a Marihuana User', pp. 141–7.

3 Becker, 'Becoming a Marihuana User' p. 145.

4 Becker, 'Becoming a Marihuana User' p. 146.

5 E. Goffman, *Frame Analysis* (Harmondsworth: Penguin, 1975), p. 13.

6 E. Goffman, *Stigma* (Harmondsworth: Penguin, 1968).

7 E. Goffman, *Asylums* (Harmondsworth: Penguin, 1968).

8 See E. Goffman, *Relations in Public* (London: Allen Lane, 1971), pp. 26–61 (on 'Territories of the Self') and particularly pp. 41–4.

9 Goffman, *The Presentation of Self in Everyday Life*, pp. 2–3.

10 Goffman, *Relations in Public*, pp. 276–7.

11 Goffman, *Frame Analysis*, p. 21.

12 For an attempt to generate a theoretical vocabulary for this task see A. Giddens, *New Rules of Sociological Method* (London: Hutchinson, 1976), pp. 120–9 and his 'Agency, Institutions and Time-Space Analysis' in K.D. Knorr-Cetina and A.V. Cicourel (eds.), *Advances in Social Theory and Methodology* (London: Routledge & Kegan Paul, 1981), pp. 161–74.

13 On the issue of wealth see the Reports of the Royal Commission on the Distribution of Income and Wealth, particularly reports 1, 4 and 5, respectively Cmnd. 6171 (1975), Cmnd. 6626 (1976) and Cmnd. 6999 (1977) (London: HMSO). For a valuable commentary see C. Hird and J. Irvine, 'The Poverty of Wealth Statistics', in J. Irvine, I. Miles and J. Evans (eds.), *Demystifying Social Statistics* (London: Pluto, 1979), pp. 190–211.

14 See E. Frankel, 'Corpuscular Optics and the Wave Theory of Light: The Science and Politics of a Revolution in Physics', *SSS* 6 (1976), pp. 141–84.

15 On the operationalisation of citation counts and other quantitative measures in science see M. Mulkay, 'Action and Belief or Scientific Discourse? A

Possible Way of Ending Intellectual Vassalage in Social Studies of Science', *PSS* 11 (1981), pp. 165–7, and G.N. Gilbert, 'Referencing as Persuasion', *SSS* 7 (1977), pp. 113–22.

16 See Ayer, *Language, Truth and Logic*, for exposition and Quine, 'Two Dogmas', in his *From a Logical Point of View* (New York: Harper and Row, 1963), for critical observations.

17 See Papineau, *For Science in the Social Sciences* (London: Macmillan, 1978), pp. 73–4.

18 A comment from an unpublished paper by J. Goldthorpe (entitled 'Causal and Interpretive Understanding in Sociology') cited in R. Wallis and S. Bruce, 'Accounting for Action: Defending the Common Sense Heresy' *Sociology*, 17 (1983), p. 106.

19 On the interpretative use of stereotypes and 'idealizations' see Goffman, *The Presentation of Self in Everday Life*, pp. 30–4.

20 Indeed, as Marsh points out in *The Survey Method* (London: Allen & Unwin, 1982), p. 3, the practitioners of statistical work are often very sensitive to the practical difficulties of statistical interpretation.

21 See M. Pollner, 'Mundane Reasoning', *PSS* 4 (1974), pp. 35–54, particularly pp. 49–53.

22 M. Pollner, '"The Very Coinage of Your Brain": The Anatomy of Reality Disjunctures', *PSS* 5 (1975), p. 427.

23 See the reasoning strategies examined in Lakatos, *Proofs and Refutations*, and in Barnes and Law, 'Whatever Should be done', *Theory and Society*, 3 (1976).

24 H. Sacks, 'Sociological Description', *Berkeley Journal of Sociology*, 8 (1963), pp. 1–19, and also his 'An Initial Investigation of the Usability of Conversational Data for Doing Sociology', in D. Sudnow (ed.), *Studies in Social Interaction* (London: Collier-Macmillan, 1972), pp. 31–74.

25 Probably the single, best-known paper is H. Sacks, E.A. Schegloff and G. Jefferson, 'A Simplest Systematics for the Organization of Turn-Taking in Conversation', *Language*, 50 (1974), pp. 696–735. A recent, comprehensive account of CA research is available in Levinson, *Pragmatics*, pp. 284–370.

26 Levinson, *Pragmatics*, p. 338.

27 See J.M. Atkinson and P. Drew, *Order in Court* (London: Macmillan, 1979), pp. 220–232. Levinson (in *Pragmatics*, pp. 322–6) also comments on the functionalist methodology of CA.

28 It must be borne in mind that Goffman's reflexivity (the interpretation of action in the light of people's presumed wants and beliefs) is not the same as the idea that Garfinkel terms reflexivity. The latter refers to the operation of the 'documentary method' (see Chapter 1).

29 A detailed account of much of the early sociology of knowledge is available in W. Stark, *The Sociology of Knowledge* (London: Routledge & Kegan Paul, 1958).

30 See A. Oakley, *Housewife* (Harmondsworth: Penguin, 1976), chapter 6, 'Four Housewives', pp. 105–55. A similar use is made of quotations in P. Willis' study of working-class culture: *Learning to Labour* (Farnborough: Gower, 1980).

Chapter 6 *Conclusion: Philosophy, method and practice*

1 Accounts of disciplines crossing this boundary also demonstrate its fragility. For an 'establishment' discipline see Wright's 'The "Success" of Medicine', and by contrast Collins and Pinch, 'The Construction of the Paranormal', both in Wallis, (ed.), *On the Margins of Science* (Keele: University of Keele, 1979).

2 P. Calvert, *The Concept of Class* (London: Hutchinson, 1982), p. 210, quoting from W.B. Gallie, 'Essentially Contested Concepts', *Proceedings of the Aristotelian Society*, 55 (1955–56), pp. 167–98.

3 Habermas, *Knowledge and Human Interests* (London: Heinemann, 1972), p. 314.

4 P.F. Strawson, *Individuals: An Essay in Descriptive Metaphysics* (London: Methuen, 1959), p. 10.

5 The 'Founding Fathers' (eg. Marx, Weber and Durkheim) are still at the centre of many undergraduate courses and text-books: for a recent study of their worth see Abrams, *Historical Sociology* (Shepton Mallet: Open Books, 1982).

6 Of course, it may prove possible to treat power as a form of distorted communication: see J. Habermas, 'On Systematically Distorted Communication', *Inquiry*, 13 (1970), pp. 205–18, and 'Hannah Arendt's Communications Concept of Power', *Social Research*, 44 (1977), pp. 3–24. For counter-arguments see A. Giddens, 'Habermas's Critique of Hermeneutics', in his *Studies in Social and Political Theory* (London: Hutchinson, 1977), pp. 135–64.

7 A small and fairly unsystematic set of examples includes the studies in J. Potter, P. Stringer and M. Wetherwell, *Social Texts and Context: Literature and Social Psychology* (London: Routledge & Kegan Paul, 1984); J. Lacan, *The Language of the Self*, with notes and commentary by A. Wilden (Baltimore: Johns Hopkins Press, 1968); and J. Mitchell, *Psychoanalysis and Feminism* (London: Allen Lane, 1974). More general commentaries include J. Coulter, *The Social Construction of Mind* (London: Macmillan, 1979), and R. Harré, *Personal Being: A Theory for Individual Psychology* (Oxford: Blackwell, 1983).

Bibliography

As in the Notes and References, two abbreviations are used: the journals *Social Studies of Science* and *Philosophy of the Social Sciences* are indicated by *SSS* and *PSS* respectively

Abrams, P. *Historical Sociology* (Shepton Mallet: Open Books, 1982)

Apel, K.O. 'Communication and the Foundations of the Humanities', *Acta Sociologica*, 15 (1972), pp. 7–27

Atkinson, J.M. and P. Drew *Order in Court* (London: Macmillan, 1979)

Atkinson, J.M. and J. Heritage (eds.), *Structures of Social Action: Studies in Conversation Analysis* (Cambridge: CUP, 1984)

Ayer, A.J. *Language, Truth and Logic* (Harmondsworth: Penguin, 1971)

Barnes, B. *Scientific Knowledge and Sociological Theory* (London: Routledge & Kegan Paul, 1974)

Barnes, B. *Interests and the Growth of Knowledge* (London: Routledge & Kegan Paul, 1977)

Barnes, B. 'On the "Hows and "Whys" of Cultural Change', *SSS* 11 (1981), pp. 481–98

Barnes, B. *T.S. Kuhn and Social Science* (London: Macmillan, 1982)

Barnes, B. and D. Edge (eds.), *Science in Context: Readings in the Sociology of Science* (Milton Keynes: Open University Press, 1982)

Barnes, B. and J. Law. 'Whatever Should be done with Indexical Expressions?', *Theory and Society*, 3 (1976), pp. 223–37

Barnes, B. and D. MacKenzie. 'On the Role of Interests in Scientific Change', in Wallis *On the Margins of Science* (Keele, University of Keele, 1979), pp. 49–66

Barnes, B. and S. Shapin (eds.), *Natural Order: Historical Studies of Scientific Culture* (Beverly Hills and London: SAGE, 1979)

Bauman, Z. *Hermeneutics and Social Science* (London: Hutchinson, 1978)

Becker, H.S. 'Becoming a Marihuana User', *American Journal of Sociology*, 59 (1954), pp. 41–58; reprinted in the Open University School and Society Course Team's *School and Society: A Sociological Reader* (London: Routledge & Kegan Paul, 1971), pp. 141–7

Benson, D. and J.A. Hughes. *The Perspective of Ethnomethodology* (London: Longman, 1983)

Bernstein, R.J. *Praxis and Action* (London: Duckworth, 1972)

Bernstein, R.J. *The Restructuring of Social and Political Theory* (Oxford: Blackwell, 1976)

Bhaskar, R. *A Realist Theory of Science*, second edn (Brighton: Harvester, 1978)

Bhaskar, R. *The Possibility of Naturalism* (Brighton: Harvester, 1979)

Bloor, D. *Knowledge and Social Imagery* (London: Routledge & Kegan Paul, 1976)

Brown, S.C. (ed.), *Philosophical Disputes in the Social Sciences* (Brighton: Harvester, 1979)

Bruce, S. and R. Wallis. 'Rescuing Motives', *The British Journal of Sociology*, 34 (1983), pp. 61–71

Calvert, P. *The Concept of Class* (London: Hutchinson, 1982)

Chalmers, A.F. *What is this Thing called Science?*, second edn (Milton Keynes: Open University Press, 1982)

Chomsky, N. *Topics in the Theory of Generative Grammar* (The Hague and Paris: Mouton, 1966)

Cicourel, A.V. *Method and Measurement in Sociology* (New York: Free Press, 1964)

Collins, H.M. 'The Seven Sexes: A Study in the Sociology of a Phenomenon, or the Replication of Experiments in Physics', *Sociology*, 9 (1975), pp. 205–24

Collins, H.M. (ed.), 'Knowledge and Controversy: Studies of Modern Natural Science' *SSS* 11 (1981), the whole of no. 1

Collins, H.M. 'The Role of the Core-Set in Modern Science: Social Contingency with Methodological Propriety in Science', *History of Science*, 19 (1981), pp. 6–19

Collins, H.M. 'The Function of Calibration in Experimental Science', paper presented to the combined meeting of HSS, PSA, SHOT and 4S, Philadelphia (1982)

Collins, H.M. (ed.), *Sociology of Scientific Knowledge: A Sourcebook* (Bath: Bath University Press, 1982)

Collins, H.M. and T.J. Pinch: 'The Construction of the Paranormal: Nothing Unscientific is Happening', in Wallis, *On the Margins of Science* (Keele: University of Keele, 1979), pp. 237–70

Collins, H.M. and T.J. Pinch. *Frames of Meaning: The Social Construction of Extraordinary Science* (London: Routledge & Kegan Paul, 1982)

Connerton, P. (ed.), *Critical Sociology* (Harmondsworth: Penguin, 1976)

Coulter, J. *The Social Construction of Mind* (London: Macmillan, 1979)

Davidson, D. *Esssays on Action and Events* (Oxford: Clarendon, 1980)

Dallmayr, F.R. and T. McCarthy (eds.), *Understanding and Social Inquiry* (London: University of Nortre Dame Press, 1977)

Edge, D. 'Quantitative Measures of Communication in Science: A Critical Review', *History of Science*, 17 (1979), pp. 102–34

Elster, J. *Ulysses and the Sirens: Studies in Rationality and Irrationality* (Cambridge: CUP, 1979)

Feyerabend, P.K. 'Consolations for the Specialist', in Lakatos and Musgrave, *Criticism and the Growth of Knowledge* (Cambridge: CUP, 1970), pp. 197–231

Feyerabend, P.K. *Against Method* (London: New Left Books, 1975)

Fleck, L. *Genesis and Development of a Scientific Fact* (London and Chicago: University of Chicago Press, 1979)

Fraassen, B.C. van, *The Scientific Image* (Oxford: Clarendon, 1980)

Gallie, W.B. 'Essentially Contested Concepts', *Proceedings of the Aristotelian Society*, 55 (1955-56), pp. 167-98

Garfinkel, H. *Studies in Ethnomethodology* (Englewood Cliffs: Prentice-Hall, 1967)

Geuss, R. *The Idea of a Critical Theory: Habermas and the Frankfurt School* (Cambridge: CUP, 1981)

Giddens, A. *New Rules of Sociological Method* (London: Hutchinson, 1976)

Giddens, A. *Studies in Social and Political Theory* (London: Hutchinson, 1977)

Giddens, A. *Central Problems in Social Theory* (London: Macmillan, 1979)

Giddens, A. 'Agency, Institutions and Time-Space Analysis', in Knorr-Cetina and Cicourel, *Advances in Social Theory and Methodology* (London: Routledge, 1981), pp. 161-74

Gilbert, G.N. and P. Abell (eds.), *Accounts and Action* (Farnborough: Gower, 1983)

Gilbert, G.N. and M. Mulkay. *Opening Pandora's Box: A Sociological Analysis of Scientists' Discourse* (Cambridge: CUP, 1984)

Goffman, E. *The Presentation of Self in Everyday Life* (London: Allen Lane, 1969)

Goffman, E. *Relations in Public* (London: Allen Lane, 1971)

Goffman, E. *Frame Analysis* (Harmondsworth: Penguin, 1975)

Goldthorpe, J. 'A Revolution in Sociology?', *Sociology*, 7 (1973), pp. 449-62

Habermas, J. *Zur Logik der Sozialwissenschaften* (Frankfurt: Suhrkamp, 1970)

Habermas, J. 'On Systematically Distorted Communication', *Inquiry*, 13 (1970), pp. 205-18

Habermas, J. *Knowledge and Human Interests* (London: Heinemann, 1972)

Habermas, J. 'A Postscript to *Knowledge and Human Interests*', *PSS* 3 (1973), pp. 157-89

Habermas, J. 'Hannah Arendt's Communications Concept of Power', *Social Research* , 44 (1977), pp. 3-24

Halfpenny, P. *Positivism and Sociology: Explaining Social Life* (London: Allen & Unwin, 1982)

Handel, W. *Ethnomethodology: How People Make Sense* (Englewood Cliffs: Prentice-Hall, 1982)

Harré, R. *The Principles of Scientific Thinking* (London: Macmillan, 1970)

Harré, R. *The Philosophies of Science* (Oxford: OUP, 1976)

Harré, R. *Personal Being: A Theory for Individual Psychology* (Oxford: Blackwell, 1983)

Hawthorn, G. *Enlightenment and Despair* (Cambridge: CUP, 1976)

Heath, A. *Rational Choice and Social Exchange* (Cambridge: CUP, 1976)

Hempel, C.G. *Aspects of Scientific Explanation and Other Essays in the Philosophy of Science* (New York: Free Press, 1965)

Hempel, C.G. *Philosophy of Natural Science* (Englewood Cliffs: Prentice-Hall, 1966)

Hesse, M.B. 'In Defence of Objectivity', *Proceedings of the British Academy*, 58 (1973), pp. 275-92

Hesse, M.B. *Revolutions and Reconstructions in the Philosophy of Science* (Brighton: Harvester), 1980)

Hindess, B. *Philosophy and Methodology in the Social Sciences* (Brighton: Harvester, 1977)

Hollis, M. 'The Limits of Irrationality', in Wilson, *Rationality* (Oxford: Blackwell, 1970), pp. 214–20.

Hollis, M. 'Reason and Ritual', in Wilson, *Rationality* (Oxford: Blackwell, 1970), pp. 221–39

Hookway, C. and P. Pettit (eds.), *Action and Interpretation* (Cambridge: CUP, 1978)

Horkheimer, M. *Critical Theory* (New York: Seabury, 1972)

Hughes, J.A. *The Philosophy of Social Research* (Harlow: Longman, 1980)

Irvine, J., I. Miles and J. Evans (eds.), *Demystifying Social Statistics* (London: Pluto, 1979)

Keat, R. and J. Urry, *Social Theory as Science*, second edn (London: Routledge & Kegal Paul, 1982)

Knorr-Cetina, K.D. *The Manufacture of Knowledge* (Oxford: Pergamon, 1981)

Knorr-Cetina, K.D. and A.V. Cicourel (eds.), *Advances in Social Theory and Methodology* (London: Routledge & Kegan Paul, 1981)

Knorr-Cetina, K.D. and M. Mulkay (eds.), *Science Observed: Perspectives on the Social Study of Science* (London and Beverly Hills: SAGE, 1983)

Kuhn, T.S. *The Structure of Scientific Revolutions*, second edn (London and Chicago: University of Chicago Press, 1970)

Kuhn, T.S. *The Essential Tension: Selected Studies in Scientific Tradition and Change* (London and Chicago: University of Chicago Press, 1977)

Lacan, J. *The Language of the Self*, with notes and commentary by A. Wilden (Baltimore: Johns Hopkins Press, 1968)

Laing, R.D. *Self and Others* (London: Tavistock, 1969)

Lakatos, *I. Proofs and Refutations* (Cambridge: CUP, 1976)

Lakatos, I. *Philosophical Papers*, 2 vols, ed. J. Worrall, and G. Currie (Cambridge: CUP, 1978) entitled: *The Methodology of Scientific Research Programmes* and *Mathematics, Science and Epistemology*

Lakatos, I. and A. Musgrave (eds.), *Criticism and the Growth of Knowledge* (Cambridge: CUP, 1970)

Larrain, J. *The Concept of Ideology* (London: Hutchinson, 1979)

Laslett, P. and W.G. Runciman (eds.), *Philosophy, Politics and Society*, III (Oxford: Blackwell, 1969)

Latour, B. and S. Woolgar. *Laboratory Life: The Social Construction of Scientific Facts* (London and Beverly Hills: SAGE, 1979)

Laudan, L. *Progress and its Problems: Towards a Theory of Scientific Growth* (London: Routledge & Kegan Paul, 1977)

Leach, E.R. *Culture and Communication* (Cambridge: CUP, 1976)

Levinson, S.C. *Pragmatics* (Cambridge: CUP, 1983)

Losee, J. *A Historical Introduction to the Philosophy of Science* (Oxford: OUP, 1980)

McCarthy, T. *The Critical Theory of Jurgen Habermas* (London: Hutchinson, 1978)

MacIntyre, A. 'The Idea of a Social Science' in Wilson, *Rationality* (Oxford: Blackwell, 1970), pp. 112-30
MacKenzie, D. 'Statistical Theory and Social Interests: A Case Study', *SSS* 8 (1978), pp. 35-83
MacKenzie, D. *Statistics in Britain, 1865-1930: The Social Construction of Scientific Knowledge* (Edinburgh: Edinburgh University Press, 1981)
Marsh, C. *The Survey Method* (London: Allen & Unwin, 1982)
Mehan, H. and H. Wood. *The Reality of Ethnometholodology* (New York and London: Wiley, 1975)
Meltzer, B.N., J.W. Petras and L.T. Reynolds. *Symbolic Interactionism: Genesis, Varieties and Criticisms* (London: Routledge & Kegan Paul, 1975)
Mitchell, J. *Psychoanalysis and Feminism* (London: Allen Lane, 1974)
Mulkay, M. *Science and the Sociology of Knowledge* (London: Allen & Unwin, 1979
Mulkay, M. and G.N. Gilbert. 'Putting Philosophy to Work: Karl Popper's Influence on Scientific Practice', *PSS* 11 (1981), pp. 389-407
Mulkay, M., J. Potter and S. Yearley. 'Why an Analysis of Scientific Discourse is Needed', in Knorr-Cetina and Mulkay, *Science Observed* (London and Beverly Hills: SAGE, 1983), pp. 171-203
Nagel, E. *The Structure of Science* (London: Routledge & Kegan Paul, 1961)
Newton-Smith, W.H. *The Rationality of Science* (London: Routledge & Kegan Paul, 1981)
Oakley, A. *Housewife* (Harmondsworth: Penguin, 1976)
Outhwaite, W. *Understanding Social Life* (London: Allen & Unwin, 1975)
Pantin, C.F.A. *The Relations Between the Sciences* (Cambridge: CUP, 1968)
Papineau, D. *For Science in the Social Sciences* (London: Macmillan, 1978)
Pinch, T.J. 'The Sun-Set: The Presentation of Certainty in Scientific Life', *SSS* 11 (1981), pp. 63-93
Pollner, M. 'Mundane Reasoning', *PSS* 4 (1974), pp. 35-54
Pollner, M. '"The Very Coinage of Your Brain": The Anatomy of Reality Disjunctures', *PSS* 5 (1975), pp. 411-30
Popper, K.R. *Conjectures and Refutations* (London: Routledge & Kegan Paul, 1963)
Popper, K. R. *The Logic of Scientific Discovery*, reprint edn (London: Hutchinson, 1980)
Potter, J., P. Stringer and M. Wetherell. *Social Texts and Context: Literature and Social Psychology* (London: Routledge & Kegan Paul, 1984)
Pratt, V. *The Philosophy of the Social Sciences* (London: Methuen, 1978)
Psathas, G. (ed.) *Everyday Language: Studies in Ethnomethodology* (New York: Irvington, 1979)
Putnam, H. *Meaning and the Moral Sciences* (London: Routledge & Kegal Paul, 1978)
Quine, W.V.O. *From a Logical Point of View* (New York: Harper and Row, 1963)
Ravetz, J.R. *Scientific Knowledge and its Social Problems* (Oxford: Clarendon, 1971)
Richards, S. *Philosophy and Sociology of Science: An Introduction* (Oxford: Blackwell, 1983)

Roemer, J.E. *Analytical Foundations of Marxian Economic Theory* (Cambridge: CUP, 1981)
Rose, H. and S. Rose (eds.), *The Political Economy of Science* (London: Macmillan, 1976)
Rudner, R.S. *Philosophy of Social Science* (Englewood Cliffs: Prentice-Hall, 1966)
Ryan, A. *The Philosophy of the Social Sciences* (London: Macmillan, 1970)
Ryan, A. (ed.), *The Philosophy of Social Explanation* (London: OUP, 1973)
Sacks, H. 'Sociological Description', *Berkeley Journal of Sociology*, 8 (1963), pp. 1–19
Sacks, H. 'An Initial Investigation of the Usability of Conversational Data for Doing Sociology', in D. Sudnow (ed.), *Studies in Social Interaction* (London: Collier-Macmillan, 1972), pp. 31–74
Sacks, H., E.A. Schegloff and G. Jefferson. 'A Simplest Systematics for the Organization of Turn-Taking in Conversation', *Language*, 50 (1974), pp. 696–735
Schwartz, H. and J. Jacobs. *Qualitative Sociology* (New York: Free Press, 1979)
Searle, J.R. *Speech Acts* (Cambridge: CUP, 1969)
Shapin, S. 'The Politics of Observation: Cerebral Anatomy and Social Interests in the Edinburgh Phrenology Disputes', in Wallis, *On the Margins of Science* (Keele: University of Keele, 1979), pp. 139–78
Skinner, Q. '"Social Meaning" and the Explanation of Social Action', in P. Laslett, W.G. Runciman and Q. Skinner (eds.), *Philosophy, Politics and Society*, IV (Oxford: Blackwell, 1972), pp. 136–57
Slack, J. 'Class Struggle Among the Molecules', in T. Pateman (ed.), *Countercourse* (Harmondsworth: Penguin, 1972), pp. 202–17
Smith, D. 'Theorizing as Ideology', in R. Turner (ed.), *Ethnomethodology* (Harmondsworth: Penguin, 1974), pp. 41–5
Strawson, P.F. *Individuals: An Essay in Descriptive Metaphysics* (London: Methuen, 1959)
Strawson, P.F. *Freedom and Resentment and Other Essays* (London: Methuen, 1974)
Suppe, F. (ed.), *The Structure of Scientific Theories* (Urbana and London: University of Illinois Press, 1977)
Symes, J. 'Policy and Maturity in Science', *Social Science Information*, 15 (1975), pp. 337–47
Taylor, C. *The Explanation of Behaviour* (London: Routledge & Kegan Paul, 1964)
Taylor, C. 'Neutrality in Political Science', in Laslett and Runciman, *Philosophy, Politics and Society*, III (Oxford: Blackwell, 1969), pp. 25–57
Thomas, D. *Naturalism and Social Science* (Cambridge: CUP, 1979)
Toulmin, S.E. *Foresight and Understanding* (London: Hutchinson, 1961)
Toulmin, S.E. *Human Understanding* (Oxford: Clarendon, 1972)
Tudor, A. *Beyond Empiricism: Philosophy of Science in Sociology* (London: Routledge & Kegan Paul, 1982)
Turner, S.P. *Sociological Explanation as Translation* (Cambridge: CUP, 1980)
Wallis, R. (ed.), *On the Margins of Science: The Social Constrution of Rejected*

Knowledge, Sociological Review Monograph 27 (Keele: University of Keele, 1979)

Wallis, R. and S. Bruce, 'Accounting for Action: Defending the Common Sense Heresy', *Sociology*, 17 (1983), pp. 97–110

Willis, P. *Learning to Labour* (Farnborough: Gower, 1980)

Wilson, B.R. (ed.), *Rationality* (Oxford: Blackwell, 1970)

Winch, P. *The Idea of a Social Science and its Relation to Philosophy* (London: Routledge & Kegan Paul, 1958)

Winch, P. 'Understanding a Primitive Society', in Wilson, *Rationality* (Oxford: Blackwell, 1970), pp. 78–111

Winner, L. 'Do Artifacts have Politics?', *Daedalus*, 109 (1980), pp. 121–36

Woolgar, S. 'Interests and Explanation in the Social Study of Science', *SSS* 11 (1981), pp. 365–94

Woolgar, S. 'Irony in the Social Study of Science', in Knorr-Cetina and Mulkay, *Science Observed* (London and Beverly Hills: SAGE, 1983), pp. 239–66

Wright, P.W.G. 'A Study in the Legitimisation of Knowledge: The "Success" of Medicine and the "Failure" of Astrology', in Wallis, *On the Margins of Science* (Keele: University of Keele, 1979), pp. 85–101

Yearley, S. 'Textual Persuasion: The Role of Social Accounting in the Construction of Scientific Arguments', *PSS* 11 (1981), pp. 409–35

Yearley, S. 'The Relationship Between Epistemological and Sociological Cognitive Interests', *Studies in the History and Philosophy of Science*, 13 (1982), pp. 353–88

Young, R.M. 'Evolutionary Biology and Ideology – then and now', *Science Studies*, 1 (1971), pp. 177–206

Index